Cell·

The Practical Approach Series

SERIES EDITORS
D. RICKWOOD
Department of Biology, University of Essex
Wivenhoe Park, Colchester, Essex CO4 3SQ, UK

B. D. HAMES
Department of Biochemistry and Molecular Biology,
University of Leeds, Leeds LS2 9JT, UK

Affinity Chromatography
Anaerobic Microbiology
Animal Cell Culture (2nd Edition)
Animal Virus Pathogenesis
Antibodies I and II
Biochemical Toxicology
Biological Membranes
Biomechanics—Materials
Biomechanics—Structures and Systems
Biosensors
Carbohydrate Analysis
Cell–Cell Interactions
Cell Growth and Division
Cellular Calcium
Cellular Neurobiology
Centrifugation (2nd Edition)
Clinical Immunology
Computers in Microbiology
Crystallization of Nucleic Acids and Proteins
Cytokines
The Cytoskeleton
Diagnostic Molecular Pathology I and II
Directed Mutagenesis
DNA Cloning I, II, and III
Drosophila
Electron Microscopy in Biology
Electron Microscopy in Molecular Biology
Enzyme Assays
Essential Molecular Biology I and II
Experimental Neuroanatomy
Fermentation
Flow Cytometry
Gel Electrophoresis of Nucleic Acids (2nd Edition)
Gel Electrophoresis of Proteins (2nd Edition)
Genome Analysis
Haemopoiesis
HPLC of Macromolecules
HPLC of Small Molecules
Human Cytogenetics I and II (2nd Edition)
Human Genetic Diseases
Immobilised Cells and Enzymes

In Situ Hybridization
Iodinated Density Gradient Media
Light Microscopy in Biology
Lipid Analysis
Lipid Modification of Proteins
Lipoprotein Analysis
Liposomes
Lymphocytes
Mammalian Cell Biotechnology
Mammalian Development
Medical Bacteriology
Medical Mycology
Microcomputers in Biochemistry
Microcomputers in Biology
Microcomputers in Physiology
Mitochondria
Molecular Genetic Analysis of Populations
Molecular Neurobiology
Molecular Plant Pathology I and II
Monitoring Neuronal Activity
Mutagenicity Testing
Neural Transplantation
Neurochemistry
Neuronal Cell Lines
Nucleic Acid and Protein Sequence Analysis
Nucleic Acids Hybridisation
Nucleic Acids Sequencing
Oligonucleotides and Analogues
Oligonucleotide Synthesis
PCR
Peptide Hormone Action
Peptide Hormone Secretion
Photosynthesis: Energy Transduction
Plant Cell Culture
Plant Molecular Biology
Plasmids
Pollination Ecology
Postimplantation Mammalian Embryos
Preparative Centrifugation
Prostaglandins and Related Substances
Protein Architecture
Protein Engineering
Protein Function
Protein Purification Applications
Protein Purification Methods
Protein Sequencing
Protein Structure
Protein Targeting
Proteolytic Enzymes
Radioisotopes in Biology
Receptor Biochemistry
Receptor–Effector Coupling
Receptor–Ligand Interactions
Ribosomes and Protein Synthesis
Signal Transduction
Solid Phase Peptide Synthesis
Spectrophotometry and Spectrofluorimetry
Steroid Hormones
Teratocarcinomas and Embryonic Stem Cells
Transcription and Translation
Virology
Yeast

Cell–Cell Interactions

A Practical Approach

Edited by
BRUCE R. STEVENSON
WARREN J. GALLIN
and
DAVID L. PAUL

This book belongs to

Oxford University Press, Walton Street, Oxford OX2 6DP
Oxford   New York   Toronto
Delhi   Bombay   Calcutta   Madras   Karachi
Kuala Lumpur   Singapore   Hong Kong   Tokyo
Nairobi   Dar es Salaam   Cape Town
Melbourne   Auckland   Madrid
and associated companies in
Berlin   Ibadan

Oxford is a trade mark of Oxford University Press

A Practical Approach 🛈 is a registered trade mark
of the Chancellor, Masters, and Scholars of the University of Oxford
trading as Oxford University Press

Published in the United States
by Oxford University Press Inc., New York

© Oxford University Press, 1992

All rights reserved. No part of this publication may be
reproduced, stored in a retrieval system, or transmitted, in any
form or by any means, without the prior permission in writing of Oxford
University Press. Within the UK, exceptions are allowed in respect of any
fair dealing for the purpose of research or private study, or criticism or
review, as permitted under the Copyright, Designs and Patents Act, 1988, or
in the case of reprographic reproduction in accordance with the terms of
licences issued by the Copyright Licensing Agency. Enquiries concerning
reproduction outside those terms and in other countries should be sent to
the Rights Department, Oxford University Press, at the address above.

This book is sold subject to the condition that it shall not,
by way of trade or otherwise, be lent, re-sold, hired out, or otherwise
circulated without the publisher's prior consent in any form of binding
or cover other than that in which it is published and without a similar
condition including this condition being imposed
on the subsequent purchaser.

Users of books in the Practical Approach Series are advised that prudent
laboratory safety procedures should be followed at all times. Oxford
University Press makes no representation, express or implied, in respect of
the accuracy of the material set forth in books in this series and cannot
accept any legal responsibility or liability for any errors or omissions
that may be made.

A catalogue record for this book is available from the British Library

Library of Congress Cataloging in Publication Data
Cell-cell interactions : a practical approach / edited by Bruce R.
Stevenson, Warren J. Gallin, and David L. Paul.
(The Practical approach series)
Includes bibliographical references and index.
1. Cell interaction.   2. Cell junctions.   3. Cell adhesion.
I. Stevenson, Bruce Russell.   II. Gallin, Warren J.   III. Paul,
David Louis.   IV. Series.
[DNLM: 1. Cell Adhesion Molecules.   2. Cell Communication.
3. Intercellular Junctions. QH 604.2. C3938]
QH604.2.C4415   1993   574.87—dc20   92–1559
ISBN 0–19–963319–3 (hbk)
ISBN 0–19–963318–5 (pbk)

Set by Footnote Graphics, Warminster, Wilts
Printed in Great Britain by Information Press Ltd, Eynsham, Oxford
Cover illustration by Jaclyn Peebles

Preface

Cell–cell interactions comprise a complex and dynamic form of communication. This communication between a cell and its immediate environment defines, to a large extent, the phenotype of that cell; it also determines the structure and function of the tissues and organs in which the cell resides. Cell–cell interactions can be mediated by several different mechanisms. Families of cell adhesion molecules are found in unique distributions on all cell types and are believed to act in developmental recognition and tissue formation. Cell adhesion molecules also play a role in the maintenance of tissue integrity. Intercellular junctions are architecturally differentiated areas of the plasma membrane that serve a variety of functions, including adhesion, intercellular communication, and occlusion of the extracellular space. This book is a collection of techniques used to study direct cell–cell and cell–substratum interactions in a variety of systems.

The methods used to study the interactions of a cell with its immediate environment are diverse, employing strategies derived from various disciplines. Our intent in assembling this volume was to choose approaches that are widely used or represent recent advances in methodology, while avoiding overt repetition of general techniques that are more commonly used in biological research. However, it is impossible to write a current techniques-oriented text that does not overlap with other fields; therefore, some of the protocols described here have been covered previously in other texts from this series. Conversely, many of the more specific techniques that are detailed in this volume have applicability in other areas of work. Although our selection of topics obviously represents our biased view of the field of cell–cell interactions, we believe we have produced a representative and useful compilation of topics that will be helpful to scientists who wish to investigate the mechanisms and consequences of cell–cell interactions in their system of choice.

We would like to thank all of the authors for their outstanding efforts, and Oxford University Press for the patience and support shown to us during the generation of this book.

Edmonton and Boston BRUCE R. STEVENSON
March 1992 WARREN J. GALLIN
DAVID L. PAUL

Contents

List of contributors xvii
Abbreviations xix

1. Assays of cell adhesion 1
S. Hoffman

1. Introduction 1
   Terminology 1
   Principles of adhesion assays 2
   Implication of specific molecules in adhesion 3
2. Preparation of reagents 4
   Cells 4
   Plasma membranes 7
   Adhesion molecules 9
3. Adhesion assays 14
   Aggregation assays 15
   Binding assays 21

References 29

2. Generation and characterization of function-blocking anti-cell adhesion molecule antibodies 31
Ann Acheson and Warren J. Gallin

1. Introduction 31
2. Monoclonal antibodies 32
   Stages of hybridoma production 32
   Dose and form of antigen 33
   Route of inoculation 35
   Collecting mouse serum from a tail bleed 35
   Developing a screening procedure 36
   Pooling for screening 39
   Expanding and freezing 39
   Single-cell cloning by limiting dilution 39
   Ascites fluid 39
   Purifying IgG from ascites 40
3. Polyclonal antibodies 41
   Immunoaffinity purification 41

Contents

Preparing the antigen for injection	44
Serum treatment	46
Fab' fragments	47
4. Testing the ability of the IgG or Fab' fragments to block function	48
Functional assays	49
5. Recombinant bacteriophage antibody libraries	51
References	53

3. Expression cloning: transient expression and rescue from mammalian cells   55
Alejandro Aruffo

1. Introduction	55
2. Overview of the cloning procedure	55
3. Vectors	57
4. Methods included in this chapter	58
5. Propagation of *SupF* plasmids	58
6. mRNA preparation	59
7. cDNA synthesis and library preparation	62
8. Transient expression screening of the cDNA library	67
DEAE–dextran transfection	68
Panning	69
Spheroplast fusion	71
9. Conclusion	73
Acknowledgements	74
References	74

4. Functional analysis of uvomorulin by site-directed mutagenesis and gene transfer into eukaryotic cells   75
Jorg Stappert and Rolf Kemler

1. Introduction	75
2. Background	75
3. General aspects	77
Site-directed mutagenesis by complementary oligonucleotides	77
Single point mutation by PCR	79
Generating mutations within a gene	81
Deletion mutants at the 3'-end of a gene	84

4. Introduction of recombinant vectors into mammalian cells	84
Transfection of coprecipitates of calcium phosphate and DNA	86
DNA transfection by electroporation	88
Lipofection of lipopolyamine-coated plasmids	88
References	89

5. Integrin mutagenesis   91
A. Reszka and A. Horwitz

1. Introduction	91
2. General approach	92
Generation of the mutation	93
Expression vectors	96
Host cell choice	97
Introduction of the cDNA into the host cells	98
Isolation and screening for expression	99
3. Applications of integrin mutagenesis	103
Focal adhesion localization of mutant integrins	103
Assay of integrin mutants for cellular adhesion	107
Activation of human $\alpha_{IIb}\beta_3$ deletion mutants	109
References	109

6. Isolation of intercellular junctions   111
Elliot L. Hertzberg, Bruce R. Stevenson, Kathleen J. Green, and Shoichiro Tsukita

1. Introduction	111
2. Rodent liver gap junctions	111
Distribution and function of gap junctions	111
Isolation principles	113
Isolation of rat liver plasma membranes	113
Isolation of rat liver gap junctions	115
Characterization of components of rodent liver gap junctions	119
Utility of preparation and future directions	119
3. Tight junctions	119
Introduction	119
Isolation of a tight junction-enriched fraction from mouse liver	121
Modifications, variations, and future directions	123
4. Desmosomes	125
Introduction	125
Isolation principles	126
Isolation of enriched preparations of desmosomes from bovine tongue	127

Technical modifications of standard protocol	129
Further fractionation of bovine tongue desmosomes	130
Characterization of desmosomal components	130
Future directions	133

5. Cell-to-cell adherens junctions — 133
- Introduction — 133
- Isolation principles — 134
- Isolation of adherens junctions from rat liver — 134
- Technical modifications — 136
- Characterization of components of the isolated AJ — 136
- Summary — 139

Acknowledgements — 139

References — 139

7. The Xenopus oocyte cell–cell channel assay for functional analysis of gap junction proteins — 143
Gerhard Dahl

1. Introduction — 143

2. The paired Xenopus oocyte expression assay — 144
- Frogs: handling and injection of HCG — 144
- Preparation of oocytes — 145
- Synthesis of mRNA — 147
- Purification of RNA — 149
- Microinjection into oocytes — 150
- Removal of the vitelline envelope — 151
- Doping and pairing — 153

3. Functional tests for expression of cell–cell channels — 155
- Dual voltage-clamp — 155
- Tracer flux — 158

4. Independent verification of expression — 158
- Protein chemistry — 158
- Immunohistochemistry — 158
- Electron microscopy — 160

5. Discriminating between endogenous and exogenous connexins — 161
- To achieve high ratios of exogenous to endogenous coupling — 161
- Time-pairing to avoid endogenous coupling — 162
- Eliminating endogenous connexins — 162
- Selective inhibition of endogenous coupling — 163

Acknowledgements — 164

References — 164

8. Patch-clamp analysis of gap junctional currents 167
R. D. Veenstra and P. R. Brink

1. Introduction: dual whole-cell patch-clamp analysis of gap junctional currents 167
2. Preparation of cell pairs 168
   Achieving the dual whole-cell recording (DWCR) configuration 168
3. Equivalent circuit for the dual whole-cell configuration 170
   Derivation of equations for junctional conductance measurements 171
4. Analysis of junctional currents 176
   Ideal preparation 176
   Effects of seal resistance on junctional resistance measurements 179
   Effects of membrane resistance on junctional resistance measurements 181
   Effects of series resistance on junctional resistance measurements 181
   Summary of effects of non-junctional resistance changes on R_j measurements 183
   Effects of changes in junctional resistance 185
   Detection of gap junction channel currents 187
5. Double whole-cell vs. single patch approach 188
6. Achieving direct cytoplasmic access to gap junction channel 190
   Optimal ionic conditions 191
   Avoiding intracellular membranes 192
   How to know if one has gap junction channels in a patch 192
7. Channel analysis: sampling methods 196
   Simple analytical approaches 197

References 199

9. Biochemical approaches for analysing *de novo* assembly of epithelial junctional components 203
Manijeh Pasdar

1. Introduction 203
2. Establishment of cultures for *de novo* assembly of desmosomes 205
3. *De novo* assembly studies 206
   General techniques 206
   Solubility properties of protein components of the junctional complex 212
   Fate of the newly synthesized proteins following metabolic labelling 212

       Intracellular processing and transport of newly synthesized
          membrane core glycoproteins   215
  4. Conclusion   220
  Acknowledgements   221
  Appendix   221
  References   225

10. Biochemical methods for studying supramolecular complexes involving cell adhesion molecules, integral membrane proteins, and the cytoskeleton   227
W. James Nelson, Rachel Wilson, and Robert W. Mays

  1. Introduction   227
       General considerations   228
  2. Cells   229
       Cell culture   229
       Induction of cell–cell adhesion in monolayer of epithelial cells   230
       Monitoring cell–cell adhesion and confluency in monolayers
          of MDCK cells growing on filter inserts   232
       Labelling membrane proteins on different cell-surface domains   234
  3. Preliminary analysis of the cellular organization of
      proteins prior to complex isolation   235
       Subcellular localization of proteins   236
       Extraction of proteins from cells   238
  4. Further separation of protein complexes   245
       Immunoprecipitation of protein complexes   247
       Non-denaturing polyacrylamide gel electrophoresis   249
  5. Conclusion: cross-referencing the results from the
      subcellular localization, extraction, and fractionation of
      proteins   253
  Acknowledgements   254
  References   254

11. Electrophysiological assessment of epithelia   257
Karl J. Karnaky, Jr.

  1. Introduction   257
       Measuring electrical properties of epithelia   258

Contents

2.	The short-circuit current technique	259
	Measuring transepithelial resistance	261
3.	Major instruments of the short-circuit current technique	262
	The voltage-clamp	262
	The Ussing chamber: a practical description of the device	263
4.	Using the chart recorder	266
5.	Trouble-shooting	267
6.	Other problems	268
	Poor mixing	268
	Noisy electrical connections	268
7.	Ion flux studies	268
	Introduction	268
	Sample calculation of isotopic flow from *Protocol* 5 data	270
	Calculation of the conversion factor	271
	Conversion of $\mu Eq/h/cm^2$ to $\mu amp/cm^2$	271
	Calculation of specific activity	271
8.	Electrophysiological studies of the paracellular pathway	272
	Isotope fluxes measured under open-circuit conditions	272
	Isotope fluxes measured at pre-selected voltages	272
	Application of the vibrating probe	272
	Flux measurements for non-ionic solutes	273
Acknowledgements		273
References		273

12. Compartmented culture analysis of nerve growth 275
Robert B. Campenot

1.	Introduction	275
	Experimental capabilities of compartmented cultures	276
2.	Construction of compartmented cultures	277
	Standard cultures	278
	Side-plated cultures	279
	Plating the neurones	282
	Pin rake	285
	Teflon dividers	287
	Silicone grease syringe	289
3.	Culture medium	291
	Methylcellulose	291
	Rat serum and ascorbic acid	291
	Effectiveness of the seal between compartments	293

4. Applications of compartmented cultures		294
Neuritotomy and collection of neurites		294
Measurement of neurite extension		295
Electrical stimulation		297
References		298

Appendix

Suppliers of specialist items — 299

Index

301

Contributors

ANN ACHESON
Regeneron Pharmaceuticals, Inc., 777 Old Sawmill River Road, Tarrytown, NY 10591–6707, USA.

ALEJANDRO ARUFFO
Bristol-Meyers Squibb Pharmaceutical Research Institute, 3005 First Avenue, Seattle, Washington 98121, USA.

P. R. BRINK
Department of Physiology and Biophysics, Health Science Center, SUNY Stonybrook, Stonybrook, NY 11794, USA.

ROBERT B. CAMPENOT
Department of Anatomy and Cell Biology, University of Alberta, Edmonton, Alberta T6G 2H7, Canada.

GERHARD DAHL
Department of Physiology and Biophysics, University of Miami Medical School, PO Box 016430, Miami, FL 33101, USA.

WARREN J. GALLIN
Department of Zoology, University of Alberta, Edmonton, Alberta T6G 2E9, Canada.

KATHLEEN J. GREEN
Department of Pathology, Northwestern University Medical School, 303 East Chicago Avenue, Ward Building, Chicago, IL 60611, USA.

ELLIOT L. HERTZBERG
Department of Neuroscience, Albert Einstein College of Medicine, 1300 Morris Park Avenue, Bronx, NY 10461, USA.

S. HOFFMAN
Medical University of South Carolina, Department of Medicine, Division of Rheumatology and Immunology, 171 Ashley Avenue, 912CSB, Charleston, SC 29425, USA.

A. HORWITZ
Department of Cell/Structural Biology, 506 Morrill Hall, University of Illinois, Urbana, IL 61801, USA.

KARL J. KARNAKY, Jr.
Department of Cell Biology and Anatomy and the Marine Biomedical and Environmental Sciences Program, Medical University of South Carolina, 171 Ashley Avenue, Charleston, SC 29425, USA.

Contributors

And The Mount Desert Island Biological Laboratory, Salsbury Cove, ME 04672, USA.

ROLF KEMLER
Max-Planck-Institut für Immunobiologie, Stubeweg 51, D-7800, Freiburg, Germany.

ROBERT W. MAYS
Department of Molecular and Cellular Physiology, Stanford University Medical Center, Stanford, CA 94305-5426, USA.

W. JAMES NELSON
Department of Molecular and Cellular Physiology, Stanford University Medical Center, Stanford, CA 94305-5426, USA.

MANIJEH PASDAR
Department of Anatomy and Cell Biology, University of Alberta, Edmonton, Alberta T6G 2H7, Canada.

DAVID L. PAUL
Department of Neurobiology, Harvard Medical School, Boston MA 02115, USA.

A, RESZKA
Department of Cell/Structural Biology, 506 Morrill Hall, University of Illinois, Urbana, IL 61801, USA.

JORG STAPPERT
Max-Planck-Institut für Immunobiologie, Stubeweg 51, D-7800 Freiburg, Germany.

BRUCE R. STEVENSON
Department of Anatomy and Cell Biology, University of Alberta, Edmonton, Alberta T6G 2H7, Canada.

SHOICHIRO TSUKITA
Department of Information Physiology, National Institute for Physiological Sciences, Myodaiji, Okazaki, 444 Japan.

R. D. VEENSTRA
Department of Pharmacology, SUNY Health Sciences Center at Syracuse, Syracuse, NY 13210, USA.

RACHEL WILSON
Department of Chemistry, Philadelphia College of Textile and Science, Philadelphia, Pennsylvania, USA.

Abbreviations

AJ	adherens junctions
ATP	adenosine triphosphate
BSA	bovine serum albumin
CAMs	cell adhesion molecules
CSK	cytoskeleton extraction buffer
CTP	cytidine triphosphate
DEAE	diethylaminoethyl
DFBS	dialysed fetal bovine serum
DG/dg	desmoglein
DMEM	Dulbecco's modified Eagle's medium
DMSO	dimethylsulfoxide
dNTP	deoxynucleotide triphosphate
DOC	sodium deoxycholate
DP/dp	desmoplakin
DPBS	Dulbecco's phosphate-buffered saline
DTT	dithiothreitol
DWCR	double whole-cell recording
ECM	extracellular matrix
EDTA	ethylenediaminetetraacetic acid
EGTA	ethyleneglycobis-(β-aminoethyl) ether N,N,N',N' tetraacetic acid
ELISA	enzyme-linked immunoabsorbent assay
EM	electron microscope
endo-H	endoglycosidase H
EtOH	ethanol (ethyl alcohol)
Fab'	immunoglobulin antigen binding fragment
FACS	fluorescence activated cell sorter
FBS	fetal bovine serum
FCS	fetal calf serum
FITC	fluoroscein isothiocynate
GTP	guanosine triphosphate
HBSS	Hanks' balanced salt solution
HCM	high Ca^{2+} concentration (1.8 mM) DMEM
HCM-chase	HCM containing 2 × methionine
HS-buffer	high stringency immunoprecipitation buffer
ICAM	intercellular cell adhesion molecule
Ig	immunoglobin
IgG	immunoglobin G
IgM	immunoglobin M

Abbreviations

IP	intraperitoneal
KLH	keyhole limpet haemocyanin
LCM	low Ca^{2+} concentration (5 μm) medium
LCM-chase	LCM containing 2 × methionine
LCM-MET	methionine-free LCM
mAb	monoclonal antibody
MDCK	Madin–Darby canine kidney (cell)
MW	molecular weight
NCAM	neural cell adhesion molecule
NGF	nerve growth factor
OD	optical density
PAGE	polyacrylamide gel electrophoresis
PBS	phosphate-buffered saline
PEG	polyethylene glycol
PMSF	phenylmethylsulfonyl fluoride
PPO	2,5-diphenyloxazole
PVC	polyvinyl chloride
SAMs	cell–substrate adhesion molecules
SC	subcutaneous
SDS	sodium dodecyl sulfate
TE	Tris–EDTA
TEA	tetraethylammonium chloride
TEMED	N,N,N',N'-tetramethylethylenediamine
TPCK	L-1-tosylamide-2-phenylethyl chloromethyl ketone
UTP	uridine triphosphate
UV	ultraviolet

1

Assays of cell adhesion

S. HOFFMAN

1. Introduction

Classic embryological studies suggested that cell adhesion along with cell migration, cell proliferation, cell death, and differentiation are the primary cellular processes of development. Later studies demonstrated that cells prepared from embryonic tissues would reaggregate in a relatively tissue-specific manner, but little was known about the molecular mechanisms involved. The study of cell adhesion became a prominent topic in modern cell biological research in the 1970s and 1980s with the identification of specific adhesion molecules. It then became possible to study and compare the functions, distributions, and structures of these molecules (1–5).

1.1 Terminology

To understand the methods presented in this chapter, it is necessary for the reader to be familiar with a few terms describing adhesion molecules. Adhesion molecules fall into two major classes, those involved in cell-to-cell adhesion and those involved in cell-to-extracellular matrix adhesion (also known as cell-to-substrate adhesion). Cell–cell adhesion molecules (CAMs) are usually integral membrane glycoproteins, although some variant forms of CAMs that arise through alternative mRNA splicing are linked to the plasma membrane by covalent bonds to the lipid phosphatidylinositol. Cell–substrate adhesion is mediated by the interaction of certain extracellular matrix proteins, glycoproteins, and proteoglycans (known as cell–substrate adhesion molecules or SAMs) with their cell-surface receptors. The most well-characterized group of cell-surface receptors for SAMs are known as integrins (see Chapter 5). Even these few classifications described here are not absolute. For example, the SAM fibronectin can promote cell–cell adhesion by simultaneously interacting with its cell-surface receptor on two adjacent cells. In another interesting exception, certain integrins have been found to mediate cell–cell adhesion by binding to CAMs (6, 7) in addition to their more typical ability to mediate cell–substrate adhesion by binding to a SAM.

Functional assays have revealed that certain CAMs have Ca^{2+}-dependent adhesion mechanisms, while other CAMs have no known requirements for particular ions. Essentially, all known SAMs require Ca^{2+}, Mg^{2+}, or Mn^{2+} for activity. Adhesion mechanisms also differ in the identity of the specific cell types and molecules involved. Adhesion between two cells from the same tissue or cell line is called homotypic while adhesion between two distinct cell types is called heterotypic. Similarly, adhesion mediated through the interaction of copies of the same protein on two adjacent cells is called homophilic, while adhesion mediated by the interaction of two distinct proteins is called heterophilic.

1.2 Principles of adhesion assays

There are two basic types of adhesion assays:

- Aggregation assays in which individual particles in a suspension of particles of relatively uniform size aggregate to form larger aggregates.
- Binding assays in which small particles bind to large particles or surfaces and can be separated from the unbound small particles by simple physical means.

The most commonly used aggregation assays and binding assays and the methods used to quantitate adhesion in these assays are summarized in *Table 1*. The physical principle behind aggregation assays is simply that, provided the particles being studied are relatively uniform in size at the beginning of the experiment, the rate of aggregation of cells or other particles in suspension during an incubation can be quantitated in terms of the changes that occur in the distribution of the sizes of the particles present. These changes in particle size distribution can be quantitated best using a particle counter.

Table 1. Common adhesion assays

Assays	Quantitation
Aggregation of:	
Cells	Particle counting
Plasma membrane fragments	Microscopy
Liposomes bearing CAMs	
Beads bearing CAMs or SAMs	
Binding of:	
Plasma membrane fragments to cell	Microscopy or quantitative dye-binding for unlabelled cells
Liposomes bearing CAMs to cells	
Beads bearing CAMs or SAMs to cells	Fluorescence microscopy or counting of radioactivity for labelled probes
Soluble CAMs or SAMs to cells	
Cells to culture dishes containing cell monolayers	
Cells to culture dishes coated with CAMs or SAMs	

When the changes are great, they can also be quantitated by microscopy if necessary.

Binding assays take advantage of the difference in size between the smaller 'probe' particles and the larger entities to which they are binding. Plasma membrane fragments, beads, liposomes, and soluble proteins that are bound to cells are separated from the unbound particles by differential centrifugation. Similarly, probe cells bound to culture dishes containing cell monolayers or protein coats are separated from unbound probe cells by a gentle washing of the culture dish. Bound probes are then detected and quantitated by microscopy, fluorescence microscopy, or scintillation or gamma counting. How these principles apply to specific assays is presented in Section 3.

1.3 Implication of specific molecules in adhesion

Given the large amount of information now available on specific adhesion molecules, most researchers who study cell adhesion phenomena will be interested in relating the adhesion they observe to a specific molecule. Methods used to implicate specific molecules in adhesion are:

- antibody inhibition
- specific peptide or oligosaccharide inhibition
- transfection and other molecular biological techniques
- binding of purified adhesion molecules to cells

In antibody inhibition experiments intact IgGs or IgMs or their monovalent Fab' fragments are added to the experimental system. They bind to the specific adhesion molecule against which they were prepared and thereby block its function. Provided that the antibodies were prepared against a sufficiently pure antigen and that they do not cross-react with other proteins, they will have no effect on any other molecule in the cell. Therefore, if an antibody inhibits adhesion while control antibodies do not, then the molecule against which the antibody was made is likely to function as an adhesion molecule in that system.

In specific peptide or oligosaccharide inhibition experiments, peptides or oligosaccharides found in a particular adhesion molecule are either isolated from digests of the molecule or synthesized. If these fragments of adhesion molecules are part of an active site, they may act as competitive inhibitors of adhesion. Specific inhibition of adhesion by such fragments implicates the molecule from which they were derived as being involved in adhesion in the system being studied.

In transfection experiments, cells that do not express a particular adhesion molecule are made to take up cDNAs that encode the molecule. Transfected cells that stably express the molecule are selected and their adhesive properties are compared to the parental cell line. In principle, this procedure alters the cells only in their expression of one protein. Therefore, if the cells become adhesive after transfection, the transfected protein is considered to be an adhesion molecule. Other genetic methods such as homologous recombination or the addition of antisense RNA have the potential to block the expression of a particular protein; however, these methods have so far been used little in the study of adhesion molcules.

Purified adhesion molecules prepared under non-denaturing conditions frequently retain the adhesive function and specificity that they had on the cell surface or in the extracellular matrix. If the binding of a putative adhesion molecule in solution to its cell surface receptor is weak, binding may be enhanced by adsorbing the protein to the surface of liposomes or beads or to a culture dish and using these multivalent ligands in a binding assay. This is likely to be the most legitimate method of comparing the function of an adhesion molecule *in vitro* to its function *in vivo*. Adhesion molecules *in vivo* are effectively multivalent because they are present in large numbers of copies in close proximity to each other either on a cell surface or in an insoluble array in the extracellular matrix.

2. Preparation of reagents

In this section, the preparation and labelling of the entities used in most adhesion assays is described. These include cells, plasma membranes, purified adhesion molecules, and adhesion molecules embedded in liposomes or attached to surfaces. The preparation of antibodies which are used to determine the molecular specificity of adhesive interactions is described in Chapter 2. There are several ways to prepare each of these reagents. Typical protocols are presented that make the principles behind the preparations clear. The reader who is interested in preparing these reagents in a particular experimental system should refer to the literature to learn the pertinent adaptations of these methods.

2.1 Cells

Cells for adhesion assays are almost always prepared from embryonic tissues or early postnatal tissues because it is difficult to release cells in a viable form from most tissues in older animals. There are two major considerations in preparing cells for adhesion assays:

- Are conditions severe enough to produce a suspension of single cells?
- Are conditions gentle enough to preserve the integrity of cell surface proteins that may function in adhesion?

Cells are frequently released from culture dishes for passage by treatments such as 0.25% crude trypsin in the presence of 1 mM ethylenediaminetetraacetic acid (EDTA). This sort of treatment is likely to be too harsh for the preparation of cells that will be immediately competent for use in adhesion assays. For the release of cells from tissue, it is reasonable to use 0.002 to 0.01% highly purified trypsin in the presence of 1 mM EDTA in preliminary studies. However, if a Ca^{2+}-dependent adhesion mechanism is present, it may be necessary to eliminate the EDTA or even to add 1 mM Ca^{2+} to preserve the structural integrity and function of the adhesion molecules. One should not expect to find conditions that will release all the cells from a tissue. It is better to settle for a limited yield of single cells which have a robust activity in adhesion assays than to try to totally disrupt a tissue. A typical method for cell preparation follows in *Protocol 1*.

Protocol 1. Cell preparation for aggregation assay

Materials
- trypsin (Worthington Biochemical Corp.)
- Hanks' balanced salt solution buffered with 20 mM Hepes pH 7.4 (HBSS)
- DNase I (Worthington Biochemical Corp.)
- 3.5% bovine serum albumin (BSA) in HBSS

Methods
1. Choose conditions (level of trypsin, presence of divalent cations or EDTA) for the release of single cells (see text for considerations and possible ranges).
2. Dissect embryonic tissue into an appropriate serum-free medium, such as HBSS either formulated containing Ca^{2+} and Mg^{2+} ions or Ca^{2+}- and Mg^{2+}-free and supplemented with 1 mM EDTA. To wash the tissue, replace the medium. Cultured cells are washed several times with the appropriate medium and trypsinized either after being scraped up using a rubber policeman, or directly on the culture dish.
3. Incubate tissue or cultured cells with gentle agitation at room temperature (RT) or 37°C under the chosen conditions, the volume of medium should be at least five times the tissue volume. After 5 min incubation, break up tissues to a limited extent by trituration using a Pasteur pipette that has been broken off to enlarge the bore and fire-polished. At any time during the protocol that cells or tissue fragments appear to be agglutinated by DNA released from broken cells, add or re-add DNase I (20 μg/ml per addition).
4. Continue to incubate the tissue or cells for an additional 15 to 30 min, occasionally triturating with increasingly narrow-bore fire-polished pipettes.
5. Harvest the tissue by centrifugation for 3 min at 700 g in a table-top centrifuge. Pellets are firmer and more easily handled when collected in a 15-ml tube rather than a 50-ml tube.

Protocol 1. *Continued*

6. All further steps should be done using ice-cold media. To remove subcellular debris, resuspend the pellet in 4 ml medium and carefully layer the suspension on top of a cushion of 7 ml of medium containing 3.5% BSA. Centrifuge 3 min at 700 g and discard the supernatant.
7. To remove cell aggregates, resuspend the pellet in 4 ml of medium and centrifuge for 20 sec at 25 g. Save the supernatant and discard the pellet.
8. To wash and harvest the cells, dilute to 15 ml with medium and centrifuge 3 min at 700 g. Repeat. Resuspend the cells in an appropriately small volume of medium. Obtain a cell count using a particle counter or a haemocytometer prior to using cells in an adhesion assay.

Some published protocols appear to be very gentle because tissues are disrupted with collagenase (which should have no effect on cell-surface proteins) and little or no trypsin. These methods should be evaluated with care, however, because some commercial collagenase preparations may also contain other proteases. To minimize the potential for unwanted proteolysis, large amounts of BSA are sometimes added during the collagenase digestion step.

Cultured cells are frequently easier to harvest as a single-cell suspension than cells from tissues because the bonds that need to be disrupted are more accessible. In some cases, treatment with EDTA alone is sufficient (8, 9). As with tissues, preliminary experiments can be performed with 0.002 to 0.01% trypsin and it may be necessary to include Ca^{2+} ions to protect Ca^{2+}-dependent adhesion mechanisms.

If the cells prepared in a pilot experiment are inactive in adhesion assays, it may still be possible to produce active cells by using gentler conditions (for example, cutting back on the level of trypsin present, eliminating EDTA if present or adding Ca^{2+}, or decreasing the duration of proteolysis). If all attempts at cell preparation fail, it may still be possible to evaluate the adhesive properties of cells from tissues or cultures by preparing plasma membrane fragments (Section 2.2) and studying their adhesive properties (see Section 3.1.2).

It is, of course, much easier to prepare cells that do not require tissue disassociation such as lymphocytes. In this case, fractionation of different cell types becomes an issue. Lymphocytes can be fractionated on the basis of their density and their expression of cell surface antigens. Antibodies can be used in positive selections to isolate cells bearing a particular antigen or in negative selection by complement-mediated lysis. For a detailed description of such methods, the reader should refer to *Lymphocytes: A Practical Approach*.

2.1.2 Labelling of cells

To detect probe cells in binding experiments, it may be necessary to label the cells prior to use. In such cases, it is critical that the label not affect the adhesive properties of the cells.

Frequently used labelling methods are:

(a) Radioactive labels
  - Uptake of ^{51}Cr into cytoplasm
  - Intrinsic label with [^{3}H]-leucine or [^{35}S]-methionine
(b) Fluorescent labels
  - Uptake of diacetyl fluorescein into cytoplasm
  - Labelling of lipids with diI (red fluorescence) or diO (green fluorescence) (available from Molecular Probes)

All of these methods are easy to use. Single-cell suspensions are labelled with ^{51}Cr (7), diacetyl fluorescein (10), or diI and diO (11) by incubating the cells with the reagent, then washing out the unbound reagent. Intrinsic radioactive label is best done by incubating tissues or cell culture overnight with the labelled amino acid in media that contains little or none of the unlabelled amino acid (12). A single-cell suspension is then prepared from the labelled tissue by normal methods.

For experiments that involve a small number of data points, it may be preferable to avoid the use of radioactivity and analyse the binding of fluorescently labelled cells by fluorescence microscopy. This method will also allow the experimenter to distinguish binding events involving a probe cell aggregate from binding events involving a single probe cell. For experiments involving a large number of samples, radioactively labelled cells may be preferable because many data points can be processed and counted in a short period of time, while analysis of the same experiment by fluorescence microscopy would be impractically arduous.

Many adhesion experiments do not require cell labelling. Cells for use in aggregation assays do not need to be labelled except in applications in which the heterotypic adhesion of two distinct cell types (coaggregation) is being monitored by fluorescence microscopy (see Section 3.1.3). Unlabelled cells are also used in binding experiments in which they are incubated with smaller labelled probe particles.

2.2 Plasma membranes

Plasma membranes are prepared by rupturing cells under conditions which avoid trapping subcellular organelles in the plasma membrane fragments and which allow the ready separation of the plasma membrane fragments from other organelles. Plasma membranes can be rapidly prepared in high purity and yield from neural tissues because neural cells have a high ratio of surface area to cytoplasm. A typical method for the preparation of plasma membranes from neural tissue is set out in *Protocol 2*.

Protocol 2. Plasma membrane purification

Materials
- PBS
- protease inhibitors

    PMSF (200 mM stock in ethanol, use a final 1 mM)
    aprotinin (50 µg/ml stock in phosphate-buffered saline (PBS), use a final 1 µg/ml)
    N-ethylmaleimide (250 mM stock in PBS, use a final 10 mM)
    benzamidine (125 mM stock in PBS, use a final 5 mM)
    leupeptin (1.25 mg/ml stock in PBS, use a final 50 µg/ml)
    pepstatin A (1 mg/ml stock in ethanol, use a final 5 µg/ml)
- 70% sucrose in water
- 42% sucrose in PBS

Methods
1. All steps should be performed with ice-cold solutions. Dissect the tissue into PBS. Wash the tissue with changes of PBS and resuspend the tissue in an equal volume of PBS.
2. Add protease inhibitors. It may not be necessary to use all the protease inhibitors listed above.
3. Disrupt the tissue, using an ultrasonic homogenizer. Adjust the homogenate to a final 10% sucrose using the 70% sucrose solution.
4. Layer the homogenate over a cushion of 42% sucrose dissolved in PBS in either a fixed angle or swinging bucket ultracentrifuge rotor. Centrifuge 30 min at 40 000 g. Nuclei and connective tissue will centrifuge to the bottom of the tubes while myelin, if present, will stay at the top. Plasma membranes will collect at the interface between the homogenate and the 42% sucrose.
5. To wash the plasma membranes, dilute with 10 vol. of PBS, pellet by centrifugation at 30 000 g for 15 min and repeat. Resuspend the membranes in PBS at a concentration consistent with further plans.

In contrast, it is more difficult to prepare pure plasma membranes from other cell types, particularly ones with a large amount of cytoplasmic membranes in the Golgi apparatus and endoplasmic reticulum. In these cases, cells are usually broken under hypotonic conditions (to swell the cells) and plasma membranes are fractionated from other membranes and organelles by equilibrium density fractionation on continuous or step sucrose gradients (13, 14). The reader should refer to the literature for the specialized details of plasma membrane purification from other cells and tissues.

Labelled plasma membrane fragments can be prepared from tissues and cells labelled intrinsically with radioactive amino acids or fluorescently with dyes such as diI or diO as described above. In addition, plasma membrane fragments can be labelled fluorescently by the inclusion of a soluble fluorescent dye such as 6-carboxyfluorescein during tissue disruption (15). The soluble dye is trapped inside plasma membrane fragments as they reseal, untrapped dye is removed when the membranes are washed.

2.3 Adhesion molecules
2.3.1 Purification
Adhesion molecules can be difficult to purify because:

- Transmembrane proteins require the presence of detergents during extraction and later chromatographic purification steps to avoid hydrophobic interactions with other molecules.
- Extracellular matrix proteins may be difficult to extract, have limited solubility, and have a tendency to interact specifically with other matrix proteins and non-specifically with a variety of proteins.
- With the exception of fibronectin and collagen, most adhesion molecules are present at low concentrations, requiring that they be prepared from large quantities of tissue.

CAMs and SAM receptors such as integrins are transmembrane proteins. To solubilize these proteins while retaining adhesive function, membranes are extracted with neutral detergents such as octylthioglucoside, NP-40, or Triton X-100, or with zwitterionic detergents. Charged detergents such as sodium dodecyl sulfate (SDS) or deoxycholate tend to denature proteins and should be avoided. Standard chromatographic methods frequently do not provide good resolution in the presence of detergents. Therefore, when possible, transmembrane adhesion molecules are fractionated by affinity chromatography. Many CAMs have been isolated by immunoaffinity chromatography using monoclonal antibodies covalently attached to beads (see Chapter 2). Many integrins have been isolated using particular extracellular matrix proteins covalently attached to beads as a support for affinity chromatography.

Protocol 3. Purification of fibronectin receptors from placenta

Materials
- cell-binding fragment of fibronectin (Telios Pharmaceuticals)
- CNBr-activated Sepharose 4B (Pharmacia-LKB Biotechnology)
- 50 mM octylthioglucoside/PBS/1 mM $CaCl_2$/1 mM $MgCl_2$/1 mM PMSF
- 25 mM octylthioglucoside/PBS/1 mM $CaCl_2$/1 mM $MgCl_2$

Protocol 3. *Continued*
- 25 mM octylthioglucoside/PBS/10 mM EDTA
- 25 mM octylthioglucoside/PBS/1 mM $CaCl_2$/1 mM $MgCl_2$/1 mg/ml Gly-Arg-Gly-Asp-Ser-Pro peptide (peptide from Calbiochem)
- wheat germ agglutinin coupled to Sepharose (Pharmacia-LKB Biotechnology)
- 0.1% NP-40/PBS/1 mM $CaCl_2$/1 mM $MgCl_2$
- 0.1% NP-40/PBS/1 mM $CaCl_2$/1 mM $MgCl_2$/200 mM N-acetylglucosamine

Methods

1. Prepare an affinity column by coupling the 120 kd cell-binding fragment of fibronectin to CNBr-activated Sepharose 4B (10 mg of the protein per millilitre of beads) following the manufacturer's instructions. The fibronectin fragment can be purchased or prepared in the laboratory (16).
2. Mix equal volumes of frozen placenta and 50 mM octylthioglucoside/PBS/1 mM $CaCl_2$/1 mM $MgCl_2$/1 mM PMSF. Allow the placenta to thaw and clarify the extract by ultracentrifugation.
3. Apply the extract to the beads in a column (20 ml of extract per millilitre of beads). Wash the beads with 5 vol. of 25 mM octylthioglucoside/PBS/1 mM $CaCl_2$/1 mM $MgCl_2$.
4. Elute the fibronectin receptor from the beads with 2 vol. of either 25 mM octylthioglucoside/PBS/10 mM EDTA (the EDTA chelates the divalent cations necessary for the interaction between fibronectin and its receptor) or 25 mM octylthioglucoside/PBS/1 mM $CaCl_2$/1 mM $MgCl_2$/1 mg per millilitre of the peptide Gly-Arg-Gly-Asp-Ser-Pro (the peptide is part of the active site of fibronectin and competitively inhibits the interaction between fibronectin and its receptor).
5. The fibronectin receptor is further purified using wheat germ agglutinin–Sepharose which binds specific carbohydrate residues found on the receptor. The sample and the beads are equilibrated in 0.1% NP-40/PBS/1 mM $CaCl_2$/1 mM $MgCl_2$ and the sample is applied to a column containing a small amounts of beads (1 ml or less). The beads are washed with several volumes of the equilibration buffer and the fibronectin receptor is eluted from the beads with two volumes of equilibration buffer supplemented with 200 mM N-acetylglucosamine.

One problem with purifications based on affinity chromatography is that frequently greater than 50% of the molecule of interest will fail to bind to the support even during repeated applications. If no affinity-based method is available for the purification of a transmembrane adhesion molecule, it may be useful to purify the extracellular portion of the molecule by standard chromatographic methods (see *Protein Purification Methods: A Practical Approach*) after first releasing the molecule from the plasma membrane by controlled proteolysis (13, 15, 17). Finally, adhesion molecules for which cDNA clones have been isolated have been prepared for use in adhesion assays using expression vectors (18).

The critical step in the purification of extracellular matrix proteins from tissues is their initial extraction. Relatively harsh conditions such as dilute acid, high salt, or concentrated urea or guanidine solutions may be required to release particular proteins. Following their solubilization, standard chromatographic techniques may be used in their purification; however, it may be necessary to maintain urea or guanidine in the buffers to keep the proteins soluble and to block their interactions with other proteins.

In some cases, extracellular matrix proteins are found to be soluble without special treatment. Fibronectin and vitronectin are present in serum (although the cellular form of fibronectin does require 1 M urea for solubilization). Homogenates of brain in neutral, isotonic buffers contain extracellular matrix proteins [such as cytotactin (also known as tenascin) and cytotactin-binding proteoglycan (CTB proteoglycan)] in solution (19). Conditioned media from cultured cells contains certain extracellular matrix proteins secreted by the cells. These proteins are, of course, mixed with any serum proteins added to the cultures. Therefore, in purifying extracellular matrix proteins from conditioned media, it may be important to limit the concentration of serum in the medium.

2.3.2 Coupling and labelling

Purified adhesion molecules are used in assays either in soluble form, embedded in liposomes, adsorbed to particles such as beads or erythrocytes, or adsorbed to culture dishes.

When the detergent in a detergent–lipid mixture is removed, the lipid forms spherical bilayers known as micelles or liposomes. If transmembrane proteins that contain lipophilic domains are present in the detergent–lipid mixture, they are likely to be incorporated into the liposomes (20, 21) (although there is no guarantee that their extracellular portion will also face outward in the liposomes). Proteins without lipophilic domains will not be incorporated into the lipid bilayer. Such reconstituted membrane systems have often been used to demonstrate the activity of proteins that form specific transmembrane transport channels. Typical methods for the preparation of liposomes are presented in *Protocol 4*.

Protocol 4. Preparation of liposomes

Materials
- phosphatidylcholine
- phosphatidylserine

Assays of cell adhesion

Protocol 4. *Continued*

- 0.5% NP-40/PBS or 25 mM octylthioglucoside/PBS
- biobeads SM-2 (Bio-Rad)
- sucrose
- 30% sucrose in PBS
- 10% sucrose in PBS
- PBS

Methods

1. Phosphatidylcholine or a 5:1 phosphatidylcholine:phosphatidylserine mixture is dried on to the wall of a Corex tube under a stream of N_2 and dissolved in a detergent solution (either 0.5% NP-40/PBS or 25 mM octylthioglucoside/PBS) using a bath sonicator to disperse the sample.
2. An adhesion molecule in the same detergent is added to the solution (the mass ratio of protein to lipid should be between 1:5 and 1:30).
3. Detergent is removed causing lipid bilayers to form including proteins with lipophilic domains. NP-40 is removed by treatment with Biobeads SM-2 (see Chapter 2, *Protocol 5*), octylthioglucoside is removed by dialysis.
4. Liposomes are fractionated away from unincorporated protein on the basis of their lower buoyant density. The liposome suspension is made 45% (w/v) in sucrose and placed in the bottom of an ultracentrifuge tube for a swinging-bucket rotor. The sample is carefully overlaid sequentially with layers of 30% sucrose in PBS, 10% sucrose in PBS, and PBS and centrifuged at high speed for 1–18 h at 4 °C. The liposomes are recovered as a white layer at the interface between the 0 and 10% sucrose and contain 50 to 90% of the input protein.

Proteins can become associated with surfaces either through covalent interactions involving their reactive groups (amino, carboxyl, hydroxyl, sulfhydryl) or through non-specific interactions between the protein and hydrophobic surfaces such as plastic. In the easiest systems to use, protein solutions are simply incubated with a surface. A percentage of the protein becomes associated with the surface either through a chemical reaction with a reactive group on the surface or by a hydrophobic interaction. The unbound protein is washed away from the surface and any remaining reactive sites on the surface are blocked by incubation with bovine serum albumin (BSA). If these simple methods fail to adsorb an adhesion molecule to a surface or if they adsorb an adhesion molecule to a surface but inactivate it, then it may be worthwhile to use more complex chemical methods to attach the protein to the surface.

Two easy methods to attach proteins to surfaces are described in *Protocol 5*. Covaspheres (*Protocol 5*, A) are microscopic beads available in several sizes in the 0.3 to 1.0 micron diameter range and containing a green fluorescent dye, a red fluorescent dye, or no dye. They have reactive chemical groups on their surface that covalently bind added protein.

Plastic surfaces (*Protocol 5*, B) for adhesion experiments include beads (22) and a variety of different size and shape culture dishes and microtitre wells. While some investigators coat tissue-culture treated plastic with adhesion molecules, it is better to coat untreated plastic (Petri dishes) because adsorption of protein is better and because cells bind to tissue culture treated plastic even without the addition of adhesion molecules.

Protocol 5. Coating surfaces with proteins

Materials
- Covaspheres (Duke Scientific) or plastic culture dishes (not tissue-culture treated)
- 10 mg/ml BSA in PBS with and without 10 mM NaN_3

Methods

A. *Covaspheres*

1. Incubate a 100 µl aliquot of Covaspheres (as provided) with 10 to 100 µg of protein in PBS for 1 h at RT. Because reactive groups on these beads bind covalently to proteins, the presence of NP-40 in the sample will not inhibit coupling. To quantitate the extent of coupling, a portion of the added protein should be radioiodinated prior to coupling.

2. To free Covaspheres from unbound proteins and to block unused reactive sites, dilute the Covaspheres to 1 ml using 10 mg/ml BSA in PBS and centrifuge 5 min at 10 000 g in a microfuge.

3. Discard the supernatant, resuspend the bead pellet in 1 ml 10 mg/ml BSA in PBS, using a bath sonicator to disperse the beads if necessary. Re-spin as above. Resuspend the pellet in 1 ml of 10 mg/ml BSA in PBS/10 mM NaN_3.

4. To quantitate coupled protein, count the final sample in a gamma counter and normalize to the known specific radioactivity and protein concentration in the original sample. For many proteins, greater than 90% of the original protein will have coupled to the beads. For other proteins (notably CTB proteoglycan) as little as 20% of the protein may couple to the beads.

B. *Polystyrene plastic surfaces (Petri dishes)*

1. Incubate the surface to be coated with a 1 to 100 µg/ml solution of protein in PBS for 1 h at RT. Because proteins adsorb to plastic surfaces through hydrophobic interactions; detergent in protein solutions will inhibit adsorption. To quantitate adsorption, a portion of the added protein should be radioiodinated.

Protocol 5. *Continued*

2. To block further adsorption, remove the sample and replace with 10 mg/ml BSA in PBS. Change the BSA solution three times rapidly, then let the final change remain with the surface for 1 h at RT.
3. To quantitate adsorbed protein, cut out and count all or a portion of a coated surface and normalize to the known specific radioactivity and protein concentration in the labelled protein sample. Efficiency of adsorption can vary from about 33% of input counts for fibronectin down to about 1% for CTB proteoglycan.

Labelling adhesion molecules allows the quantitation of the protein concentration in liposomes or adsorbed to surfaces as well as the quantitation of the binding to cells of adhesion molecules in soluble form, in liposomes, or adsorbed to particles. It is important to be able to quantitate the adsorption of adhesion molecules to surfaces because these molecules can differ greater than tenfold in the fraction of input protein that adsorbs to plastic surfaces. This factor must be taken into consideration when normalizing and comparing adhesion data obtained with different molecules.

The most highly labelled molecules can be prepared by radioiodination following purification, either using Chloramine T or lactoperoxidase to catalyse the covalent attachment of ^{125}I to tyrosine residues (see *Radioisotopes in Biology: A Practical Approach*). If radioiodination inactivates a protein, it may be useful to label the protein intrinsically, then purify it or to label the protein by reductive methylation following its purification (see *Radioisotopes in Biology: A Practical Approach*). Liposomes may also be labelled by the inclusion of radioactive lipids or by including the soluble fluorescent dye, 6-carboxyfluorescein, during the formation of the liposomes. The dye becomes trapped in the interior spaces of the liposomes. Covaspheres are fluorescent; therefore, these particles are readily detected by fluorescence microscopy when they are coated with adhesion molecules and bound to cells.

3. Adhesion assays

Adhesion assays, like any other experiment, require appropriate controls because of the danger of obtaining false-positive results. For example, occasionally cells may aggregate non-specifically due to the presence of DNA released by damaged cells. The most universally useful specificity control in adhesion assays is antibody inhibition, i.e. adhesion should occur in the presence of monovalent Fab' fragments prepared from the IgG from unimmunized rabbits but should be inhibited by Fab' fragments prepared from the IgG from rabbits immunized with an adhesion molecule active in the system being studied.

3.1 Aggregation assays

During the course of aggregation assays, particles such as cells, plasma membrane fragments, liposomes, or beads form aggregates. To monitor the time-course of aggregation, an automatic particle counter is required. One source of popular and reliable models of particle counters is Coulter Electronics (Hialeah, Florida). In these machines, particles are detected as they interrupt an electric field across an aperture through which the sample is passing. Each particle interrupts the electric field with a magnitude related to its size, but counts as only one interruption event. The machine counts the number of electric field interruptions in a sample that are larger than a threshold chosen by the investigator. To obtain a histogram of particle sizes in a sample, the investigator can manually count the same sample several times using different thresholds or can purchase an accessory known as a Channelyzer which automatically produces a histogram from a single run.

Aggregation assays can be interpreted most reliably when the original suspension of particles is uniform in size. For cell aggregation, uniformity is achieved by taking care to use a single-cell suspension. It is most convenient in cell aggregation experiments to set the particle counter's threshold so that single cells are counted and to monitor aggregation in terms of the disappearance of single cells that occurs as they form aggregates (23). For smaller particles, uniform suspensions are achieved either by incubation in a bath sonicator (preferable for beads) or by passing the sample through a submicron filter (preferable for plasma membrane fragments and liposomes). Because particle counters do not reliably detect these submicron particles, it is most convenient to set the threshold so that the monomeric particles in the original suspension are not counted while the aggregates that form during an incubation are counted (20).

The major advantages and disadvantages of various aggregation assays are presented in *Table 2*. Basically, cell aggregation is the most straightforward assay, can be used to analyse large numbers of samples, but may be weak and difficult to detect in some systems. Plasma membrane aggregation can be used on tissues from which cells are difficult to prepare (for example, adult brain; ref. 20) and ensures that the molecular form and concentration of CAMs is not altered by the trypsinization that occurs during cell preparation. Liposome and bead aggregation are useful for implicating specific molecules in adhesion. Their use is not limited to homophilic adhesion mechanisms. A heterophilic interaction can be demonstrated by the ability of particles bearing one ligand to aggregate with particles bearing a second ligand provided that neither batch of particles aggregates by itself.

3.1.1 Cell aggregation

Methods for cell aggregation are presented in *Protocol 6* and sample data that might be obtained in such an experiment are presented in *Table 3*. Note that

Assays of cell adhesion

Table 2. Advantages and disadvantages of aggregation assays

Cell aggregation	
Advantages	Suitable for a large number of samples.
Disadvantages	Sometimes specific aggregation is weak relative to non-specific effects
Plasma membrane aggregation	
Advantages	Good for tissues from which cells cannot be prepared (e.g. most adult tissues).
	CAMs retain *in vivo* structure and concentration.
	One preparation can be used for many experiments on many days.
Disadvantages	Technically difficult to analyse many samples in an experiment, especially without a Channelyzer.
Liposome aggregation	
Advantages	Provide direct implication of specific CAMs in adhesion and their binding mechanisms.
	One preparation can be used for many experiments on many days.
Disadvantages	Technically difficult to analyse many samples in an experiment, especially without a Channelyzer.
	Possible functionally critical interactions with cytoskeleton or other cell-surface proteins in the same cell are lost.
	Inappropriate hydrophobic interactions between transmembrane proteins may alter their orientations and functions.
Bead aggregation	
Advantages	Provide direct implication of specific CAMs in adhesion and their binding mechanisms.
	Also useful for demonstrating intermolecular interactions between components of the extracellular matrix.
	One preparation can be used for many experiments on many days.
Disadvantages	Possible functionally critical interactions with cytoskeleton or other cell-surface proteins in the same cell are lost.
	Inappropriate hydrophobic interactions between transmembrane proteins may alter their orientations and functions.

the decrease in particle number in the control samples (non-immune Fab') is much greater than in the presence of an antibody that inhibits adhesion (50% vs. 20% at 20 min and 80% vs. 30% at 40 min). The decrease in particle number in the sample containing immune Fab' may be due to adhesion mechanisms not blocked by this antibody, to non-specific adhesion, or to non-

Table 3. Sample cell aggregation data

	Time-points				
	0 min	20 min	Δ%	40 min	Δ%
1. Non-immune Fab'	12 500	6250	50	2500	80
2. Immune Fab'	12 500	10 000	20	8750	30
3. Immune Fab' + antigen	12 500	7750	38	5000	60

The large numbers are the particles counted by a Coulter Counter in 1/40 of the 0.2-ml aliquot taken at each time-point from a scintillation vial that contained 5×10^6 cells in 2 ml at time 0. Δ% is the % change in particle number that has occurred in a sample since time 0.

---

specific cell loss such as on to the wall of the scintillation vial. Because of these effects, antibody inhibition by an immune Fab' must decrease the Δ% by at least 20 to be considered good evidence that the protein used as the immunogen is involved in adhesion in the cells being studied.

Protocol 6. Cell aggregation

Materials
- SME medium (Eagle's minimum essential medium with spinner salts buffered with 20 mM sodium phosphate or Hepes pH 7.4 and containing 20 µg/ml of DNase)
- immune Fab' in PBS
- non-immune Fab' in PBS
- scintillation vials
- 1 ml aliquots of 1% (v/v) glutaraldehyde in PBS

Methods
1. Cells (5×10^6 neural cells or 2×10^6 liver cells or other large cells) in 50 µl SME medium are pre-incubated on ice for 15 min with 0.1–0.5 ml of PBS containing either 0.5–3 mg of non-immune Fab', 0.5–3 mg of immune Fab' intended to inhibit aggregation, or any other reagent being tested for its effect on aggregation.
2. On a carefully timed basis, aggregation is initiated by diluting each sample to 2 ml with SME medium, transferring it into a glass screw-cap scintillation vial (28 mm diameter), and incubating the vial at 37°C and 90 r.p.m. on a rotary shaker. If necessary, the assay can be scaled down by using smaller vials and proportionally smaller volumes.

Protocol 6. Continued

3. At 0 min and at intervals such as 20 min, 40 min, and 60 min, remove 200 μl of sample from each tube and mix it with 1 ml of 1% glutaraldehyde in PBS. It is critical that the interval between the initiation of aggregation (time of transfer of each sample to a scintillation vial) and the harvesting of each time-point is as similar as possible for each sample.

4. Count and record the number of particles at each time-point, and calculate the percentage decrease in particle number (aggregation) for each sample over time.

Antibodies that inhibit adhesion may be prepared against either purified adhesion molecules or total cells or cell extracts (see Chapter 2). If antibodies against a purified adhesion molecule are pre-incubated with that molecule, or if antibodies against whole cells (which may recognize 1000 proteins) are pre-incubated with a single, purified adhesion molecule present on those cells, the ability of the antibody to inhibit adhesion may be reversed or 'neutralized'. In the example in *Table 3*, line 3, the degree of neutralization is 60% because pre-incubation with antigen has removed 60% of the antibody's ability to inhibit aggregation:

$$(\Delta\% \text{ line 3} - \Delta\% \text{ line 2})/(\Delta\% \text{ line 1} - \Delta\% \text{ line 2}) = 60\%.$$

This neutralization of antibody activity was used to assay the purification of several CAMs in early studies (13, 15, 24). Antibodies against whole cells were prepared and found to inhibit aggregation. These antibodies were then incubated with chromatographic fractions from cell, tissue, or plasma membrane extracts and the neutralization activity in these fractions was monitored. Ultimately, single proteins with neutralizing activity were purified and thereby identified as CAMs.

3.1.2 Aggregation of smaller particles

Methods for plasma membrane, liposome, and bead aggregation assays are similar and are presented in *Protocol 7*. Sample data is presented in *Table 4*. Because the particles in these assays are small and uniform at the beginning of the experiment and because a necessary step in the aggregation that then occurs is a series of collisions between two particles, a reasonable analogy can be made between second-order chemical kinetics and particle aggregation. Therefore, rate constants indicating the relative aggregation activity of a particle preparation can be calculated from the equation:

$$v = k_{agg}a^2$$

where k_{agg} is the apparent rate constant, v is the initial rate of aggregation, and a is the initial particle concentration (20). While Covasphere aggregation

Table 4. Sample liposome and Covasphere aggregation data

A. *Liposomes*

Particle size range (μm^3)   **Particle number**

Liposome preparation No. 1 incubated at 1 mg/ml of lipid

	0 min	5 min	10 min	15 min
4–8 (average 6)	100	200	200	100
8–16 (average 12)	0	0	50	80
16–32 (average 24)	0	0	0	10
Total (sum of particle number × average size in each range)	600	1200	1800	1800

Liposome preparation No. 2 incubated at 1 mg/ml of lipid

	0 min	5 min	10 min	15 min
4–8 (average 6)	25	25	50	75
8–16 (average 12)	0	0	0	0
16–32 (average 24)	0	0	0	0
Total (sum of particle number × average size in each range)	150	150	300	450

The volume of superthreshold particles in liposome preparation No. 1 increases at a constant rate between 0 and 10 min and plateaus between 10 and 15 min. Note that a different and inappropriate interpretation would result from just counting the number of particles larger than $4\,\mu m^3$. There is a lag period between 0 and 5 min before the total volume of superthreshold particles in liposome preparation No. 2 increases, a constant increase then occurs between 5 and 15 min. The plateau in the extent of aggregation of rapidly aggregating particles and the lag before the observation of aggregation in slowly aggregating particles are routine. During their periods of linear increase, the rates of aggregation are: liposome preparation No. 1, $120\,\mu m^3/min$; liposome preparation No. 2, $30\,\mu m^3/min$. This four-fold difference is similar to the difference in rate of aggregation of liposomes containing the adult and embryonic forms of NCAM (at similar concentrations in micrograms of NCAM per milligram of lipid).

B. *Covaspheres*

Protein(s) on Covaspheres	Soluble protein	Superthreshold particles
NgCAM	Non-immune Fab'	48 000
NgCAM	Anti-NgCAM Fab'	1000
Cytotactin	None	1660
CTB proteoglycan	None	420
Cytotactin, CTB proteoglycan	None	49 500

In this experiment, the number of superthreshold particles was determined only at a single time and at a single threshold. These data show that NgCAM Covaspheres aggregate suggesting strongly that NgCAM binds to NgCAM. The specificity of this aggregation is indicated by the fact that anti-NgCAM Fab' reduces the aggregation to background levels. Covaspheres coated with cytotactin or CTB proteoglycan do not aggregate, but when the two batches of Covaspheres are mixed, strong aggregation occurs. These results indicate that cytotactin binds to CTB proteoglycan. (Data adapted from refs 32 and 33.)

Assays of cell adhesion

may be analysed in this manner, in some cases the results obtained are so striking that it is only necessary to count the entire sample at one time-point at a single threshold (see *Table 4*, B). These data clearly indicate that NgCAM can bind to NgCAM and that cytotactin binds to CTB proteoglycan. While the interaction between cytotactin and CTB proteoglycan does not appear to mediate cell–cell or cell–substrate adhesion, this example shows that Covaspheres can be used to demonstrate heterophilic interactions.

Protocol 7. Plasma membrane, liposome, or bead (Covasphere) aggregation

Materials
- 0.8 μm-pore size Uni-Pore filter (Bio-Rad)
- 10 mg/ml BSA in PBS

Methods

1. Prepare a particle suspension that contains few counts on a Coulter Counter set to detect particles larger than 4 μm^3. Pass plasma membrane or liposome through a 0.8 μm-pore size Uni-Pore filter or incubate Covaspheres in a bath sonicator for 20 sec.

2. Incubate plasma membranes or liposomes in 0.5 ml 10 mg/ml BSA in PBS in small scintillation vials at 70 r.p.m. and RT. For brain plasma membranes, 10–100 μg of membrane protein is appropriate. For NCAM-containing liposomes, 0.2–2.0 mg of lipid containing 10–30 μg of NCAM per milligram of lipid is appropriate. These numbers are likely to vary considerably for plasma membrane from different tissues (25) and for different adhesion molecules. Incubate 100 μl of Covaspheres coated with a single protein or a mixture of two 50 μl aliquots each coated with an individual protein at RT. As described in *Protocol 5*, A, Covaspheres should be in the presence of 10 mg/ml BSA in PBS and at 1/10 the concentration supplied by the manufacturer.

3. At intervals, aliquots are removed and the Coulter Counter is used to prepare a histogram of aggregate sizes. This can be done manually (see *Table 4*, A) or using the Channelyzer accessory. From the histograms, the total volume of all superthreshold particles is determined and an initial rate of accumulation of superthreshold particle volume is calculated. It is critical to take a sufficient number of time-points to define the time period when the total volume of supertheshold particles is rising linearly and to calculate the rate of superthreshold particle appearance during that time-period.

3.1.3 Coaggregation and sorting out

The basic cell aggregation assay described in Section 3.1.1 provides a quantitative method for monitoring the aggregation of cells in suspension, but cannot provide a simple analysis of the ability of two distinct cell types to bind to each other. The interaction of two cell types can be studied by coaggregation (described here) or by the interaction of probe cells of one type with a monolayer of cells of a second type (see Section 3.2). In coaggregation experiments, each cell population is labelled with a distinct fluorescent dye (such as diI and diO) and the cells are coincubated. If aggregates form that contain both cell types, as detected by fluorescence microscopy, then the cell types are capable of coaggregation (25). The specificity and molecular mechanism of co-aggregation can be analysed further in antibody inhibition experiments.

It has long been known that following coaggregation, the two cell types present in a coaggregate may sort out. This phenomenon was originally observed in histological analyses of coaggregates and has more recently been observed by the segregation of dye-labelled cells (11). These studies, performed with transfected cells, indicate that sorting-out can be mediated not only by CAMs of different specificity but also by different amounts of the same CAM on the two cell populations.

3.2 Binding assays

Binding assays are technically easy to perform. In all cases it is straightforward to separate unbound probes from bound probes and to quantitate binding by microscopy or counting radioactivity. The further advantages and disadvantages of various binding assays are summarized in *Tables 5* and *7*. Binding assays are particularly appropriate for studying cell–cell adhesion between two distinct cell types. They are also the standard methodology for studying cell–substrate adhesion. The four cell–substrate adhesion assays summarized in *Table 7* are very different in terms of their particular advantages and disadvantages. It would be worthwhile for the person doing their first assay to carefully consider each of these possibilities.

3.2.1 CAM binding assays

As indicated in *Table 5*, CAM-mediated binding is detected in assays in which single cells bind to cell monolayers or CAM-coated substrates or in which smaller CAM-bearing probes bind to cells. Typical methods for the binding of single cells to cell monolayers are presented in *Protocol 8*. As in other adhesion assays, the standard specificity control in this assay is antibody inhibition. When this assay is used to measure heterotypic adhesion, an additional specificity control may be gained by changing the cell type in the monolayer. For example, neurones were found to bind well to monolayers of glial cells but not to monolayers of fibroblasts (26).

Table 5. Advantages and disadvantages of binding assays

CAM binding assays

Binding of labelled cells to a cell monolayer
Advantages   Standard assay for heterotypic adhesion between two cell types.

Binding of labelled plasma membranes to cells
Advantages   Good for heterotypic adhesion especially when cells cannot be prepared from one of the cell types.
                   CAMs retain *in vivo* structure and concentration.
                   One preparation can be used for many experiments on many days.

Binding of labelled liposomes to cells
Advantages   Provide direct implication of specific CAMs in adhesion.
                   One preparation can be used for many experiments on many days.

Binding of cells to CAM-coated substrates
Advantages   Provide direct implication of specific CAMs in adhesion.

SAM binding assays

Binding of cells to SAM-coated substrates
Advantages   Standard assay for cell-substrate adhesion (see *Table 6* for advantages and disadvantages of variations of this assay).

Binding of labelled liposomes to SAM-coated substrates
Advantages   Standard assay for the binding of a purified SAM cell-surface receptor to the SAM.
                   One preparation can be used for many experiments on many days.

SAM or CAM binding assays

Binding of labelled soluble adhesion molecules to cells
Advantages   For SAMs, Scatchard analysis provides number of receptors and binding constant.
Disadvantages   Binding data for CAMs is hard to interpret because they form aggregates through hydrophobic interactions.

Binding of labelled beads to cells
Advantages   Provide direct implication of specific CAMs and SAMs in adhesion.
                   One preparation can be used for many experiments on many days.

In all cases where transmembrane proteins are incorporated into liposomes or bound to beads or surfaces, disadvantages include the loss of possible functionally important interactions with the cytoskeleton or other cell-surface proteins in the same membrane and inappropriate hydrophobic interactions between transmembrane proteins that may alter their orientation and function.

Protocol 8. Binding of single cells to cell monolayers

Materials

- SME medium (Eagle's minimum essential medium with spinner salts buffered with 20 mM sodium phosphate or Hepes pH 7.4 and containing 20 μg/ml of DNase)
- PBS

Methods

1. Prepare a cell monolayer in a 35-mm culture dish by culturing cells to confluence.
2. Rinse the monolayer and add 5×10^6 fluorescently-labelled cells in SME medium and incubate 30 min at RT at 50 r.p.m. on a rotary shaker.
3. Remove unbound cells either by sequentially submerging the culture dish in several bowls of PBS or by gently removing and replacing the medium several times with a pipette.
4. Detect bound cells by fluorescence microscopy, count the number of bound cells in 10–20 microscopic fields in each culture dish at ×400 magnification.

While cells in suspension will bind to CAM-coated substrates, these substrates have been used more extensively to study neurite outgrowth. These studies have indicated that pre-coating a Petri dish with a solution of nitrocellulose prior to coating it with a CAM leads to an enhanced neurite outgrowth, presumably due to enhanced CAM adsorption to the nitrocellulose (27).

Methods for the binding of smaller probes to cells are presented in *Protocol 9* and sample data are presented in *Table 6*. In addition to antibody inhibition and comparison of cell types, an additional specificity control that is routinely used in these experiments is preparing probes that should be inactive such as liposomes without protein or beads coated only with BSA.

Protocol 9. Binding of smaller probes to cells

Materials

- SME medium (Eagle's minimum essential medium with spinner salts buffered with 20 mM sodium phosphate or Hepes pH 7.4 and containing 20 μg/ml of DNase)
- SME medium plus 3.5% BSA

Assays of cell adhesion

Protocol 9. *Continued*

Methods

1. Prepare and fluorescently or radioactively label plasma membrane fragments, liposomes containing a CAM, or beads bearing a CAM or a SAM. Make the probe particles uniform in size as described in *Protocol 7*.
2. Incubate aliquots of probe with $2 \times 10^6 - 1 \times 10^7$ cells in 0.3–1.0 ml of SME medium for 20 min at RT or 37°C. If necessary to reduce background binding, 0.5 mg/ml of BSA can be added or the incubation can be done on a rotary shaker at 70 r.p.m.
3. Cells are then separated from unbound probe: centrifuge the samples for 2 min at 250 g, resuspend the pellets in 1 ml SME medium, load them carefully on top of 3 ml of SME medium containing 3.5% BSA and centrifuge 3 min at 250 g.
4. Resuspend the pellets in 1 ml SME medium and quantitate probe binding by counting radioactivity or by fluorescence microscopy. Fluorescence microscopy can be scored in terms of bound particles per cell or percent of cells bearing bound particles. It may be necessary to set a minimum criterion for a positive cell, such as containing three or more discrete points of fluorescence.

Table 6. Binding of Covaspheres to cells

Cell type	Protein on Covaspheres	Fab' fragments	% Input bound
Neurones	N–CAM	Non-immune	3.8
Neurones	N–CAM	Anti-N–CAM	2.2
Neurones	BSA		0.4
Fibroblasts	N–CAM		0.5

Aliquots of Covasphere preparations coated with radioiodinated N–CAM or BSA were incubated with neurones or fibroblasts as described in *Protocol 9* in the presence of 1 mg of Fab' fragments prepared from non-immune or anti-N–CAM IgG. The percentage of input counts bound to the cells was quantitated. Anti-N–CAM inhibited the binding of N–CAM-coated Covaspheres to neurones. Binding of BSA-coated Covaspheres to neurones or N–CAM-coated Covaspheres to fibroblasts was at background levels. (Data adapted from refs 32 and 34.)

3.2.2 SAM binding assays

In SAM binding assays, cells bind to substrate coated with these extracellular matrix proteins. Various SAM binding assays differ, however, in the pattern in which SAMs are applied to the substrate and in the forces that bring cells into contact with the substrate. These parameters give the various assays very specific advantage and disadvantages (see *Table 7*). The methods for these assays are provided in *Protocol 10*. Sample data are presented in *Table 8* and *Figure 1*.

Table 7. Advantages and disadvantages of various cell-substrate adhesion assays

Gravity-driven assays

A. *SAM-coated 96-well flat-bottomed microtitre wells*
Advantages   Allows rapid quantitation of the binding of large numbers of samples when labelled cells are used.
Disadvantages   Must use labelled cells or dye-label cells after binding to allow quantitation.
Poor for microscopic examination of bound cells.

B. *Small SAM-coated dots on a 35-mm diameter culture dish*
Advantages   Good for precious samples because very small volumes of protein are used.
Good for microscopic examination of the morphology of bound cells (spread vs. round).
Cell do not need to be labelled.
Disadvantages   Scoring large numbers of samples can be tedious.

Centrifugation-driven assays

C. *SAM-coated 96-well flat-bottomed microtitre wells*
Advantages   The relative adhesivity of different cells for different substrates can be quantitatively compared.
Disadvantages   Must use radioactively labelled cells to allow quantitation.
Great care must be taken in setting up the apparatus for this assay.
Poor for microscopic examination of bound cells.

D. *SAM-coated 96-well U-shaped microtitre wells*
Advantages   Results obtained in minutes after addition of cells to coated wells.
Strong binding is detected to SAMs that inhibit cell-spreading (such as cytotactin) even though little or no binding is detected in other assays.
Disadvantages   Assay is only semi-quantitative.
Poor for microscopic examination of bound cells.

Methods are described in detail as follows: A, ref. 9; B, ref. 28; C, ref. 12; D, refs 8 and 28.

Protocol 10. SAM binding assays

(A, B, C, D refer to the description of assays in *Table 7*)

Materials

- DMEM (Dulbecco's modified Eagle's medium)
- PBS
- 3% formaldehyde in PBS

Protocol 10. Continued

- 0.5% toluidine blue
- 1% SDS

Methods

1. Coat plastic with dilutions of various SAMs, then block with BSA. For assays A, C, and D, use 50 μl of a protein solution in each microtitre well. For assay B, carefully place up to nine 3-μl dots in a circle midway between the centre and edge of a 35 mm Petri dish. Mark the position of each dot with a marking pen.

2. Remove the blocking solution and add cells in DMEM: 2×10^4 in 200 μl for A and D, 2×10^4 in a full well for C, and 2×10^5 in 2 ml for B. Cells are unlabelled in B and D and intrinsically labelled with [^{3}H]leucine in C. In A, cells can be radioactively labelled or unlabelled.

3. (a) Incubate 60 min at 37°C. Remove unbound cells by washing the wells with several changes of PBS. To detect bound cells, count the radioactivity in labelled cells or fix unlabelled cells with 3% formaldehyde and stain with 0.5% toluidine blue. Bound cells can then be counted directly or lysed with 1% SDS and quantitated by dye absorbance at 600 nm using a plate reader.

   (b) Incubate 60 min at 37°C. Remove unbound cells either by submerging the dishes sequentially in several bowls of PBS or by several changes of PBS in the dish. Fix with 3% formaldehyde and examine by phase-contrast microscopy. This assay provides the best conditions for examining the morphology of bound cells.

   (c) Invert another set of microtitre wells, also filled with medium on to the first. The wells are held together without leaks by double-stick carpet tape. Pairs of communicating wells form assay chambers. Centrifuge the assembly at 17 g for 8 min at 4°C to initiate cell contact with the coated substrate. Incubate the assemblies for 0 to 60 min at temperatures between 4 and 37°C. Invert replicate assemblies and centrifuge for 8 min at 4°C at a series of forces between 10 and 800 g. Cells that no longer adhere at a given force are centrifuged into the opposite well. Then freeze the plates in a dry ice–ethanol bath, cut off the bottom 3 mm of each pair of opposing wells, and determine the radioactivity in these samples. The data allow the quantitation of the percentage of input cells whose adhesion to a given substrate can withstand a given g force.

   (d) Centrifuge the microtitre plates immediately at 250 g for 1 min. Observe the pattern of cells in each well using dark-field microscopy and interpret in terms of a balance between centrifugal force and cell–substrate adhesion. On a non-adhesive substrate such as BSA, centrifugal force predominates and cells are driven into a pellet at the bottom of the well. On highly adhesive substrates, cells coat the entire bottom of the well, presumably because they bind where they first contact the well. Intermediate results can also be obtained in which increasing adhesivity is correlated with an increasing diameter of the area covered by cells after centrifugation.

Specificity controls for cell–substrate adhesion assays include antibodies against the SAMs, antibodies against their cell-surface receptors known as integrins, and peptides whose sequences are found in the active sites of the SAMs. The joint use of monoclonal antibodies that block integrin function and peptides is shown in *Table 8* in an experiment in which the equivalent of assay A was used. Fibronectin has multiple sites involved in cell–substrate adhesion. One of these sites contains the active sequence GRGDSP and binds to an integrin known as $\alpha 5\beta 1$ or VLA-5, another contains the active sequence LHGPEILDVPST and binds to the $\alpha 4\beta 1$ or VLA-4 integrin. In *Table 8*, anti-VLA-4 blocks the VLA-4 system but because the VLA-5 system is predominant and unblocked, adhesion is only slightly decreased. When the peptide that blocks the VLA-5 system is added along with the anti-VLA-4, then adhesion is strongly inhibited. Conversely, anti-VLA-5 strongly blocks adhesion but leaves some residual binding through the VLA-4 system. Further addition of the peptide that blocks the VLA-4 system strongly inhibits even this residual adhesion; addition of the peptide that blocks the VLA-5 system has little effect, because the VLA-5 system is already blocked.

The observation that different SAMs have different effects on cell mor-

---

Table 8. Binding of T cells to fibronectin-coated microtitre wells

Inhibitors of adhesion		% Input cells bound
Monoclonal antibodies	**Peptides**	
None	None	54
Anti-VLA-4	None	46
Anti-VLA-4	GRGDSP	19
Anti-VLA-4	LHGPEILDVPST	38
Anti-VLA-5	None	19
Anti-VLA-5	GRGDSP	15
Anti-VLA-5	LHGPEILDVPST	9

Microtitre wells were coated with 20 µg/ml of fibronectin, blocked with BSA and incubated with ^{51}Cr-labelled PMA-activated T cells in the presence of 10 µg/ml of monoclonal antibodies and 500 µg/ml of peptides. After a 1 h incubation at 4°C, plates were rapidly warmed to 37°C for 10 min, washed five times with PBS, the bound cells lysed with detergent, and gamma emissions counted to quantitate the % of input cells bound. See the text for the interpretation of these data. (Data adapted from ref. 35.)

Assays of cell adhesion

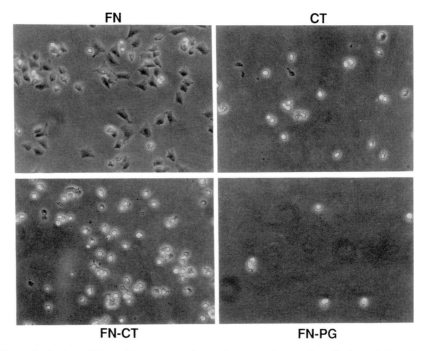

Figure 1. Endocardial cushion tissue cell attachment and spreading. Endocardial cushion tissue cells were isolated from 5-day embryonic chicken hearts and were incubated with substrates coated with fibronectin (FN), cytotactin (CT), fibronectin and cytotactin (FN-CT), or fibronectin and CTB proteoglycan (FN-PG). All proteins used to coat cell substrates were present at 100 μg/ml. Bound cells were photographed 1 h after plating. Cells attached to fibronectin spread, cells attached to cytotactin remained round. Either cytotactin or CTB proteoglycan inhibited cell attachment and spreading on fibronectin.

phology is an important topic in current research on cell–substrate adhesion. *Figure 1* shows an experiment done using assay B which provides the best optical conditions for observing cell morphology. Cells spread well on fibronectin, but remain round on cytotactin. Fibronectin is known to cause cell-spreading by providing a signal through its integrin receptors that organizes the cytoskeleton. Cytotactin must send an opposing signal to the cytoskeleton (rather than no signal) because, when fibronectin and cytotactin are mixed, cells remain round as on cytotactin alone. CTB proteoglycan appears to have an even more potent ability to inhibit cell attachment and spreading on fibronectin.

Assays C and D shed further light on the functional differences between cytotactin and fibronectin. Using assay C, it was found that cells have a comparable strength of adhesion to fibronectin and cytotactin at the time when they first contact the substrate. As time progresses, adhesion to fibronectin strengthens considerably through a process involving the cytoskeleton;

in contrast, the strength of adhesion to cytotactin decreases somewhat (12). These observations are totally consistent with results obtained with assay D, an assay in which observations are only made at the time of initial contact of cells to the substrate. In this assay, cytotactin and fibronectin had similar adhesion activities (28) (as in assay C at time 0). This is in contrast to the apparent weaker adhesion activity of cytotactin than fibronectin observed using assays A (29, 30) and B (28) (which are technically similar to assay C at later times).

When novel integrins are purified, their binding specificity can be evaluated in an assay in which the integrin is incorporated into liposomes and incubated with microtitre wells coated with various SAMs (21). Liposome preparation is as described in *Protocol 4*, using labelled lipids or radioiodinated integrins to make the liposomes radioactive. The binding assay itself is totally analogous to *Protocol 10*, assay A, except that liposomes are used instead of cells and the time of incubation of the probe with the well is increased to enhance binding.

The binding of soluble SAMs to cells can be assayed using a method analogous to *Protocol 9* for the binding of labelled small probes to cells (31). In these assays, it is critical to label the SAM in a manner that does not affect its activity and to subtract from each data point the level of non-specific binding obtained in a parallel control in which a vast excess of the unlabelled SAM is added. These 'cold competition' corrections can sometimes be larger than the specific binding. It may be impractical to do these experiments for SAMs that cannot be purified in large quantities because it may not be possible to devote enough of the SAM to the needed cold competition experiment. If saturable, dose-dependent specific binding is obtained, the data can be subjected to Scatchard analysis to determine the number of receptors per cell for the SAM and their apparent dissociation constant.

References

1. Edelman, G. M. and Crossin, K. L. (1991). *Annu. Rev. Biochem.*, **60**, 155.
2. Takeichi, M. (1990). *Annu. Rev. Biochem.*, **59**, 237.
3. Hemler, M. E. (1990). *Annu. Rev. Immunol.*, **8**, 365.
4. Yamada, K. M. (1989). *Curr. Opin. Cell Biol.*, **1**, 956.
5. Sanes, J. R. (1989). *Annu. Rev. Neurosci.*, **12**, 491.
6. Marlin, S. D. and Springer, T. A. (1987). *Cell*, **51**, 813.
7. Elices, M. J., Osborn, L., Takada, Y., Crouse, C., Luhowskyj, S., Hemler, M. E., and Lobb, R. R. (1990). *Cell*, **60**, 577.
8. Saunders, S. and Bernfield, M. (1988). *J. Cell Biol.*, **106**, 423.
9. Bourdon, M. A. and Ruoslahti, E. (1989). *J. Cell Biol.*, **108**, 1149.
10. Brackenbury, R., Rutishauser, U., and Edelman, G. M. (1981). *Proc. Natl Acad. Sci. USA*, **78**, 387.
11. Friedlander, D. R., Mège, R.-M., Cunningham, B. A., and Edelman, G. M. (1989). *Proc. Natl Acad. Sci. USA*, **86**, 7043.

12. Lotz, M. M., Burdsal, C. A., Erickson, H. P., and McClay, D. R. (1989). *J. Cell Biol.*, **109,** 1795.
13. Gallin, W. J., Edelman, G. M., and Cunningham, B. A. (1983). *Proc. Natl Acad. Sci. USA*, **80,** 1038.
14. Kartner, N., Alon, W., Swift, M., Buchwald, M., and Riordan, J. R. (1979). *J. Membr. Biol.*, **36,** 191.
15. Grumet, M. and Edelman, G. M. (1984). *J. Cell Biol.*, **98,** 1746.
16. Pytela, R., Pierschbacher, M. D., Argraves, S., Suzuki, S., and Ruoslahti, E. (1987). In *Methods in Enzymology*, Vol. 144 (ed. Leon W. Cunningham), pp. 475–89. Academic Press, Orlando, Florida.
17. Cunningham, B. A., Hoffman, S., Rutishauser, U., Hemperly, J. J., and Edelman, G. M. (1983). *Proc. Natl Acad. Sci. USA*, **80,** 3116.
18. Spring, J., Beck, K., and Chiquet-Ehrismann, R. (1989). *Cell*, **59,** 325.
19. Hoffman, S., Crossin, K. L., and Edelman, G. M. (1988). *J. Cell Biol.*, **106,** 519.
20. Hoffman, S. and Edelman, G. M. (1983). *Proc. Natl Acad. Sci. USA*, **80,** 5762.
21. Pytela, R., Pierschbacher, M. D., and Ruoslahti, E. (1985). *Cell*, **40,** 191.
22. Kadmon, G., Kowitz, A., Altevogt, P., and Schachner, M. (1990). *J. Cell Biol.*, **110,** 193.
23. Brackenbury, R., Thiery, J.-P., Rutishauser, U., and Edelman, G. M. (1977). *J. Biol. Chem.*, **252,** 6835.
24. Thiery, J.-P., Brackenbury, R., Rutishauser, U., and Edelman, G. M. (1977). *J. Biol. Chem.*, **252,** 6841.
25. Hoffman, S., Chuong, C.-M., and Edelman, G. M. (1984). *Proc. Natl Acad. Sci. USA*, **81,** 6881.
26. Grumet, M., Rutishauser, U., and Edelman, G. M. (1983). *Science*, **222,** 60.
27. Langenaur, C. and Lemmon, V. (1987). *Proc. Natl Acad. Sci. USA*, **84,** 7753.
28. Friedlander, D. R., Hoffman, S., and Edelman, G. M. (1988). *J. Cell Biol.*, **107,** 2329.
29. Erickson, H. P. and Taylor, H. C. (1987). *J. Cell Biol.*, **105,** 1387.
30. Chiquet-Ehrismann, R., Kalla, P., Pearson, C. A., Beck, K., and Chiquet, M. (1988). *Cell*, **53,** 383.
31. Akiyama, S. K. and Yamada, K. M. (1985). *J. Biol. Chem.*, **260,** 4492.
32. Hoffman, S. and Edelman, G. M. (1987). *Proc. Natl Acad. Sci. USA*, **84,** 2523.
33. Grumet, M. and Edelman, G. M. (1988). *J. Cell Biol.*, **106,** 487.
34. Grumet, M., Rutishauser, U., and Edelman, G. M. (1982). *Nature*, **295,** 693.
35. Shimizu, Y., Van Seventer, G. A., Horgan, K. J., and Shaw, S. (1990). *J. Immunol.*, **145,** 59.

2

Generation and characterization of function-blocking anti-cell adhesion molecule antibodies

ANN ACHESON and WARREN J. GALLIN

1. Introduction

Cell adhesion molecules (CAMs) are cell-surface molecules that mediate the adhesion of one cell type to another. There are two classes of CAMs defined by the ability of calcium ions to regulate their binding, calcium-dependent and calcium-independent. The neural cell adhesion molecule (NCAM) and L1 are examples of calcium-independent CAMs, and the cadherin family mediate calcium-dependent binding (see ref. 1 for review). Many CAMs are homophilic ligands, i.e. they bind to themselves, functioning as their own receptor (1).

CAMs must be localized, characterized, and their activity manipulated in order to answer questions about their biological functions in a given tissue or cell type. To characterize the presence of CAMs and to establish which forms of a given CAM are present on certain cells, antibodies which specifically recognize individual CAM molecules or portions of molecules are valuable tools. The most common approach to manipulating function is to generate function-blocking antibodies against a given CAM molecule or region of the molecule. Monoclonal antibodies that recognize a CAM often do not inhibit its function; production of more monoclonal antibodies in hope of preparing a function-blocking antibody or production of high titre polyclonal antisera are the best approaches to preparing a useful functional reagent.

In this chapter we discuss the preparation and characterization of both monoclonal and polyclonal antibodies and methods for testing the ability of these reagents to block the function of CAMs. The detailed protocols for fusion and maintenance of hybridomas are presented in *Antibodies,* Vol. I: *A Practical Approach,* and can be found in several other methods manuals (for example, ref. 4). We also provide information on the relatively new technique of preparation of permanent high diversity cloned libraries of monovalent antibody fragments in bacteriophage vectors.

2. Monoclonal antibodies

2.1 Stages of hybridoma production

The stages of hybridoma production and their approximate durations are outlined briefly below, then each step is discussed in more detail in the following sections. Monoclonal antibody (mAb) production takes three to four months and often up to a year. There is a lot of tissue culture work, so appropriate biosafety equipment must be readily available.

(a) Immunization (1 month to 1 year)
   i. Primary injection of antigen.
      Animals (almost always mice) are injected with an antigen preparation.
   ii. Boosts
      2 weeks after the primary immunization, and every 2–3 weeks thereafter, additional injections of the antigen are given to boost the immune response.
   iii. Test bleeds
      10 days after the first boost, (and also after each additional boost), a test bleed is done to determine whether the animal has developed a good humoral immune response. Usually, several boosts are needed before a high-affinity IgG response is seen.

(b) Screen development (at least 2 weeks)
   As immunizations proceed, an appropriate screening procedure is developed. Sera from the test bleeds can be used to develop and validate the screening procedure. Before immunization is even started, you should have a screening plan designed.

(c) Hybridoma production (1 month until the first screen, then at least another month for final screening).
   i. Final boost
      Several days prior to the fusion, animals are given a final boost with the antigen.
   ii. Fusion
      Antibody-producing cells from the immunized animal are mixed with myeloma cells, and then the two cell types are fused together using polyethylene glycol. Cells are then plated in a medium which selectively allows only fused cells to survive; these are the hybridoma cells. Hybridoma cells are ready for screening about 1 week after fusion.
   iii. Pooling for screening
      The best way to screen large numbers of hybridomas for positive clones is to pool many together, then identify positive pools and narrow them down to positive clones later. Positive pools and/or

clones should always be expanded and frozen before further steps are taken, to ensure that the clones are not lost. Several re-screens are needed to make sure that the positive result on the screening procedure is genuine, and that it comes from a single hybridoma clone.

2.2 Dose and form of antigen

The amount of antigen needed to produce an immune response depends on the properties of the protein being used. Some proteins or portions of proteins are highly antigenic, while other proteins produce very little immune response. In the case of the NCAM family of proteins, most antibodies generated recognize all of the NCAM forms within a species, indicating that regions of the molecule shared by all family members constitute the major antigenic sites (2). In the case of the family of cadherin molecules, most antibodies do not recognize more than one family member, indicating that antigenic epitopes differ considerably between family members (1, 3).

There are several different forms of CAMs which can be used in the immunization procedure:

(a) Purified CAMs. Inject 10–50 μg of protein per animal, depending on the quantity of protein available and its purity. More than 200 μg should usually not be injected, even if the antigen is not very pure. Purity can be determined by running the CAM preparation on an SDS gel and silver staining it to determine how many different protein bands are present in addition to the one(s) of interest.

(b) Recombinant CAMs. Either fusion proteins or full-length polypeptide chains can be produced in prokaryotic or eukaryotic expression systems. These systems have the advantage of producing large quantities of protein. However, the potential disadvantage is that fusion proteins or recombinant polypeptides may not be post-translationally modified or folded correctly when they are produced by expression systems.

(c) Synthetic peptides linked to a carrier protein. A portion of the CAM sequence of interest can be synthesized as a peptide and chemially linked to a larger carrier protein, usually keyhole limpet haemocyanin (KLH). The subsequent screening must include a means of screening out antibodies against the KLH moiety alone. One way to do this is to link the synthetic peptide to bovine serum albumin (BSA) as well as to KLH. Antibodies can then be screened against the BSA-peptide conjugate, which will effectively eliminate any immunoreactivity against KLH.

(d) Cell membrane fractions. Because CAMs are surface molecules, a good way to enrich for CAMs is to prepare a plasma membrane fraction from the cells or tissues of interest. This has at least one potential advantage over using purified CAMs; the CAM in a membrane prep is more likely to be in its native conformation than it is in solution. Another potential immunogen is living or fixed whole cells. Again, the advantage of using whole cells is that the conformation and microenvironment of the CAM is more likely to be that in which the molecule actually functions. Inject 10^5–10^7 cells per immunization.

Protocol 1. A typical immunization schedule

Materials
- antigen (amount based on factors discussed above)
- complete Freund's adjuvant
- incomplete Freund's adjuvant
- two glass syringes connected by a luer fitting
  - six female mice (Balb/c is the most commonly used strain)

Method
1. For each mouse, mix the antigen with an equal volume of complete Freund's adjuvant. This is an oil-based adjuvant which, in its complete form, contains killed *M. tuberculosis* and non-metabolizable oils (5). It is excellent for stimulating strong and prolonged immune responses. The antigen and adjuvant have to be mixed vigorously to produce an emulsion. This can be done either by extensive vortexing (for small volumes), by pushing the solutions back and forth through two glass syringes connected by a luer fitting or a three-way valve, or by using a Dounce homogenizer. Complete Freund's adjuvant is only used for the primary immunization.
2. Inject six female Balb/c mice intraperitoneally. Female mice are chosen because they tend to be easier to handle. Each individual mouse, even if genetically related, will produce a different immune response, so it is best to immunize several animals at once, to maximize the chances of getting the type of immune response desired.
3. After 14 days, repeat the injections, but use incomplete Freund's adjuvant (without *M. tuberculosis*). Boosts can be given either intraperitoneally or subcutaneously.
4. Collect tail bleeds (see Section 2.4) from mice on day 24 and do a preliminary screen to estimate the titre of the immune response. This screen should be something fast and simple, like a dot blot (described below).
5. On day 35 inject all animals with antigen in incomplete Freund's.
6. On day 45, do tail bleeds and test for immune response.
7. On day 56, give the best responder the final boost, in this case using an intravenous route without any adjuvant. Give all the other mice the same boost as was given on day 35. Multiple boosts with the same antigen will yield antibodies with higher affinity, especially when the boosts are spaced

out over weeks or months. However, multiple injections will not necessarily increase the number of epitopes that the immune response is directed towards.

8. On day 59, fuse splenocytes (using protocols from *Antibodies*, Vol. I: *A Practical Approach* or from ref. 4) from the best responder and start the procedure of hybridoma production and screening. The remaining mice can be given a final boost and fused later if something goes wrong on the first try.

2.3 Route of inoculation

There are three routes of inoculation that can be used on mice, intraperitoneal (IP), subcutaneous (SC) and intravenous (IV). The advantages and disadvantages of these routes are summarized in *Table 1*.

2.4 Collecting mouse serum from a tail bleed

Each individual mouse must be marked so that they can be identified again later. Consult your animal care facility for the recommended method of marking mice. To collect blood for the initial screens to determine whether an immune response has taken place, first warm the mouse up under an infrared lamp to increase blood flow to the tail. Swab a portion of the tail about 2 inches from the body with alcohol. Using a sterile scalpel, nick the underside of the tail across one of the visible lower veins. Collect several drops (approximately 200 to 500 µl) of blood in a tube.

To obtain the serum from the blood sample, first heat the blood to 37 °C for 1 h. Dislodge the blood clot and transfer the tube to 4 °C overnight. Spin at 10 000 g for 10 min at 4 °C. Remove the serum from the cell pellet, and spin it again to ensure that all the cells are removed. The yield should be between 50

Table 1. Routes of inoculation

Route	Primary inj. and boosts	Final boost	Maximum volume	Adjuvant	Comments
IP	Good	Fair	0.5 ml	+/−	If used for final boost, wait 5 days for fusion
SC	Good	Poor	0.2 ml	+/−	Local response; serum levels slower to increase
IV	Poor	Good	0.2 ml	No Freund's	Use soluble proteins with low salt, no detergent

and 200 μl. Sodium azide (0.02%) can be added to prevent bacterial contamination, and serum can be stored frozen at −20°C.

2.5 Developing a screening procedure

The screening procedure is the most important part of the preparation of mAbs. Since hybridoma cells all grow at approximately the same rate, all of the hybridoma supernatants will be ready to screen at about the same time. This means screening between 200 and 20 000 hybridoma colonies. The design of the screening procedure should thus have three goals:

- to identify potential positives in as short a time as possible (<48 h)
- to be easy enough to do that it can be readily carried out on as many as 20 000 samples
- to be stringent enough to reduce the number of cultures to a reasonable level manageable by one person (50 wells or less).

The two basic strategies for screening are antibody capture and antigen capture (4), which are discussed in detail below. Early in the screening process, immunoglobulin (Ig) subclass determination to enable selection for IgGs is highly recommended. The initial immune response involves production of IgMs, followed later by IgG production (4). If the fusion is done after only one or two boosts, IgMs may still be a major component of the immune response. However, working with IgM molecules is inconvenient, because IgMs are not easily cleaved into smaller fragments, and commercially available secondary antibodies to IgMs are variable in quality and are usually more expensive than those directed against IgGs. Good kits which are fast and easy to use are commercially available for Ig subclass determination (e.g. Boehringer Mannheim). Assuming that IgGs are present in a given hybridoma supernatant, the following methods can be used to screen the hybridoma library.

2.5.1 Antibody capture assays

In this type of screen, the antigen is bound to a solid phase, and the antibodies present in the hybridoma supernatant are allowed to bind to the antigen. Unbound antibody is washed away, and the bound antibody is detected by a second antibody that specifically recognizes mouse IgG and is coupled to either peroxidase or alkaline phosphatase. The enzyme reactions are developed using substrates which produce coloured products, and a 'positive' clone is identified by the presence of a coloured signal.

i. *Dot blots or 1-site ELISAs using purified or partially purified antigen*

Dot blots or enzyme-linked immunoabsorbent assays (ELISAs) are the simplest way to screen the hybridomas quickly. Small amounts of the antigen which has been used for immunization are dotted on to nitrocellulose (6) (dot blot) or added to polyvinylchloride (PVC) wells (see *Antibodies*, Vol. I: *A Practical Approach* and refs 4 and 7) (1-site ELISA). Aliquots of each different hybridoma supernatant are added to the dots, and positive clones can easily be identified using anti-mouse IgG antibodies coupled to an enzyme, followed by development of the enzyme reaction to form coloured product. In addition, different portions of the molecule of interest, either in the form of proteolytic fragments or synthetic peptides can be dotted on to nitrocellulose for screening. The ability of mAbs to recognize molecules with (or without) post-translational modifications such as phosphorylation or glycosylation can also be screened for in this manner. An additional screen which may be useful is to use the same antigen purified from different species to screen for a broad (or narrow) range of species specificity. Broad species specificity is often an advantage, especially if you might want to sell your antibody to a company for marketing.

ii. Western blots
In this type of screen, the molecules present in the antigenic material or in an appropriate tissue extract are separated by molecular mass on an SDS-polyacrylamide gel. The separated proteins are electrophoretically transferred on to nitrocellulose membrane, where they bind with high affinity (8). The membrane is cut into small strips, with each strip representing a single lane of the gel. These strips are incubated with different hybridoma supernatants, and specifically bound antibodies detected using an enzyme linked anti-mouse IgG.

Western blots have the advantage over dot blots that immunoreactivity with individual proteins or proteolytic fragments in a mixture can be identified. However, the proteins on a Western blot have been denatured and reduced by SDS-PAGE, and epitopes may only partially renature after the removal of SDS. If a given mAb binds to antigen on a Western blot, it will not necessarily recognize the native antigen.

iii. Immunostaining
In this type of screen, antibodies present in the hybridoma supernatant are bound to fixed tissue sections or cells (9) and detected using secondary antibodies coupled to fluorophores or to enzymes. The advantage of using immunostaining as a screen is that the antigen of interest can be spatially localized within a tissue. In addition, the lack of immunostaining in inappropriate cells or tissues can be taken as good evidence of the specificity of the mAb. Care must be taken to determine the level of background staining seen in the tissue by using normal mouse IgG and/or eliminating the first antibody as controls.

2.5.2 Antigen capture assays
In this type of screen, the antibodies present in the hybridoma supernatant bind to the radioactively-labelled antigen (4, 10). The antibodies are bound to

a solid phase, either PVC wells or beads, so that specifically bound antigen can easily be separated from unbound. This type of screen requires that the purified antigen be available in large enough quantities to be radioiodinated (5–10 μg).

An example of this type of assay is immunoprecipitation. This is an antigen capture assay in solution. Labelled antigen is mixed with the hybridoma supernatant in solution. Specifically bound antigen is detected by adding protein A beads, which specifically bind IgG molecules (11), thus removing the mouse IgG-labelled antigen complex from solution. For this type of screen to be successful, the antibody must have a relatively high affinity. Immunoprecipitation is usually not a practical screen for the initial stages of narrowing down positive clones, but can be very useful for detecting high affinity mAbs in later stages where several genuine positive clones have been identified.

2.5.3 Elimination of non-specific binding

In either antibody or antigen capture assays, the elimination of non-specific binding will be very important so as not to generate false/positives. Non-specific binding sites on all solid supports (nitrocellulose or polyvinyl chloride) are blocked by adding extraneous protein, commonly 1–3% BSA/phosphate-buffered saline (PBS), 0.5% casein/PBS, 2% dry skimmed milk/PBS, or 1% gelatine/PBS. Sometimes 0.1–0.3% Tween-20 is included in the blocking solution. In addition, solid phases must be washed fairly stringently between antibody incubations with low concentrations of non-ionic detergents and/or high salt concentrations to eliminate unbound antigen or antibody.

2.5.4 Limitations of monoclonal antibodies

Monoclonal antibodies recognize single epitopes, and many of those epitopes are sensitive to the conformation of the protein. Thus, a given mAb may not work on all of the different screens described above. For example, the antibody might work on a Western blot but not in immunostaining or immunoprecipitation. Thus, if you want to use your mAb for a variety of different applications, you should devise a screening protocol that tests all of the different methods you eventually will want to use.

2.5.5 Functional assay

This type of screen is the most time-consuming, since it requires that functional assays be done with each of the candidate hybridoma supernatants. For this reason, it should be left to a later stage of the overall screening procedure.

A functional assay for CAMs involves either aggregation or attachment of cells or plasma membrane fractions. These assays require preparation of relatively large quantities of single cells or partially purified plasma membrane fractions. The ability of the hybridoma supernatant to alter the rate of adhesion of these cells or membranes with each other, or their rate of attachment to a defined surface, is then assessed. Functional assays work better with Fab' fragments. Since rate measurements are involved, data are collected from several different time points (see Chapter 1). Most mAbs that have so far been developed against CAMs are not function blocking (2). This means that a large number of candidate hybridomas may have to be screened before a function-blocking mAb is found.

2.6 Pooling for screening

Pooling of medium from multiple wells in the original 96-well plates is the best way to reduce the total number of screening tests that must be done on a single day. Hybridoma supernatants can be pooled in a two-dimensional matrix, with one pool made from each row and one pool made from each column, yielding 20 pools from a 96-well plate. The correct wells can be identified by the location of intersecting positives. Once positive pools are identified, rescreening is used to confirm positive wells.

2.7 Expanding and freezing

When positive wells have been identified, the cells should be grown up in progressively larger tissue culture dishes in selection medium. Once they are growing well in a 60-mm dish, the drug selection can be removed (4, 13). Cells are then harvested, pelleted, and an aliquot frozen in liquid nitrogen, while another aliquot is used for cloning. Save the supernatant for screening.

2.8 Single-cell cloning by limiting dilution

The next step is to clone the antibody-producing cell. Single cell cloning is usually done in the presence of feeder cells or conditioned medium, because hybridoma cell-plating efficiency is low. Feeder cells can include macrophages, thymocytes or fibroblasts. The feeder cells should be placed into 96-well microtitre plates along with a fixed volume of medium. Serial dilutions of the hybridoma suspension of interest are plated out in the 96-well plate. Wells are examined for the presence of single clones after 7–10 days (see ref. 4 for further details of this protocol). When positive clones have been selected by the screening protocol, they should be re-cloned by limiting dilution at least once more, to ensure that the antibodies are indeed monoclonal.

2.9 Ascites fluid

Ascites fluid is an intraperitoneal fluid extracted from mice which have been injected intraperitoneally with hybridoma cells. The hybridoma cells continue to divide and reach high density in the peritoneal cavity, while still secreting the antibody of interest, creating a high-titre antibody solution (typically 1–10 mg/ml) which can be easily collected.

Generation and characterization of function-blocking

Protocol 2. Production of ascites fluid

Materials
- 5 adult female mice, isogenic with the originally immunized mice
- Pristane or incomplete Freund's adjuvant
- 1×10^6 hybridoma cells per mouse
- 5-ml syringe with 18-gauge needle

Method

1. Prime adult female mice with 0.5 ml pristane or incomplete Freund's adjuvant injected IP. This will cause an irritation, recruiting monocytes and lymphoid cells to the area and creating a permissive growth environment for the hybridoma cells, much like a feeder layer in culture.
2. After 1–2 weeks, inject 1×10^6 hybridoma cells per mouse in no more than 0.5 ml PBS IP.
3. Ascites fluid starts to build up within 1–2 weeks following the injection. Tap the fluid when the abdomen is noticeably distended, but before the animal has obvious trouble moving. Be aware that ascites fluid production is considered to be a painful procedure, and consult your animal care committee guide-lines before carrying out this procedure. In a sedated mouse, the ascites fluid can be drawn carefully out of the peritoneal cavity into a 5-ml syringe using an 18-gauge needle.
4. Many mice will continue to produce ascites fluid, and a second or even third batch can be collected. The animal should be sacrificed if it is in obvious distress. A single mouse can yield up to 10 ml ascites fluid per batch. Sacrifice the mouse after the last ascites collection.
5. Spin the ascites fluid at 3000 g for 10 min to separate fat (on top) and cells (pellet) from the fluid. Remove the fat layer, and carefully transfer the fluid to a new tube. Ascites fluid can be sterile-filtered and stored frozen at $-20\,°C$.

2.10 Purifying IgG from ascites

Purified IgG is useful for a number of techniques, but is crucial for functional assays. Ascites fluid is a much better source of immune IgG than tissue culture supernatant, because the concentration of antibody is many times higher. There are two commercially available systems which rapidly purify mouse IgG from ascites fluid:

   i. Pierce Immunopure kit: This kit comes complete with column, binding buffer, and elution buffer, as well as excellent directions.

   ii. Bio-Rad Affigel blue: Dialyse the ascites fluid against 20 mM Tris, pH 7.2, then follow the instructions which come with the column.

2.10.1 Monitoring the yield and purity of IgG

The elution fractions from the column are monitored by measurement of the OD_{260}, which gives an estimate of protein concentration. Protein-containing fractions are then pooled, and a protein assay can be carried out to determine the yield of IgG (usually the OD_{260} is an adequate measure of IgG concentration, assuming that 1 mg/ml has an OD_{260} of 1.4). Aliquots of the elution fractions, the starting material and the pooled fractions are run on SDS-PAGE. Stain the gel with Coomassie blue to determine purity (Note: IgGs do not stain well with silver!). Confirm the identity of the bands by staining a Western blot of the gel with anti-mouse IgG.

3. Polyclonal antibodies

Although it is possible to isolate CAMs by the standard techniques of protein chemistry (see *Protein Purification Methods: A Practical Approach*), usually the first step in characterizing a CAM is to prepare an mAb, as described above. The aim of this section is to use that antibody to purify sufficient antigen for immunizing animals to make polyclonal antibodies. This is important because often mAbs will only function in some of the many assays and procedures that you want to use. A good, monospecific polyclonal antibody will be useful for Western blotting, immunoprecipitation, immunocytochemistry and direct perturbation of the function of the CAM, while most mAbs will only work in some or even just one of these techniques.

The process of preparing a good polyclonal antibody consists of three steps:

(a) isolating the antigen;

(b) preparing the antigen for injection into the rabbit;

(c) collecting the serum and purifying IgG from the serum.

Rabbits are the most commonly used animals for producing polyclonal antibodies, but it is sometimes desirable to produce much larger amounts of serum than a rabbit can safely produce. In this case, goats or sheep are a good choice. Included below are conditions for isolating IgG from goat serum as well as rabbit serum. The only difference is the composition of the buffer used for ion-exchange adsorption.

3.1 Immunoaffinity purification
3.1.1 Preparing the column
The first step in preparing an antigen using a specific mAb is to couple the IgG to a solid support. The method below describes coupling to Sepharose CL-4B, a gel filtration resin that has a large enough exclusion size that the chances

of molecular sieving interfering with the binding of the antigen is minimal. Activated Sepharose (with complete instructions for coupling) is also available commercially (Pharmacia), and is a good choice if you are only preparing small volumes of affinity resin.

Protocol 3. Coupling mAbs to Sepharose

Materials
- purified IgG from the hybridoma of choice with the final protein concentration adjusted to 2 mg/ml by diluting with PBS to a final OD_{260} of 2.8
- cyanogen bromide (**Warning:** very toxic, must be handled in a fumehood)
- Sepharose CL-4B (Pharmacia)
- PBS

1. Weigh out 10 g cyanogen bromide (in a fume hood), transfer to a 500-ml flask, add 200 ml distilled water and a magnetic stirring bar. Seal with Parafilm and stir until completely dissolved (20–30 min).

2. Take 60 ml of packed Sepharose CL-4B beads (these beads are supplied as a slurry), resuspend in distilled water, and wash thoroughly by suction of 500 ml distilled water through the packed beads in a coarse sintered glass funnel or on a Buechner funnel with coarse filter paper. Transfer the Sepharose to a 1-litre beaker and add 150 ml distilled water. Resuspend by gently stirring with a magnetic stirrer.

3. Adjust the pH to between 11 and 11.5 with a 6 M solution of NaOH. Keep the pH electrode in the solution to monitor the pH of the suspension as the reaction progresses. Add the 200 ml of CNBr solution to the stirring suspension of beads and maintain the pH between 11 and 11.5 during the reaction period by adding drops of 6 M NaOH. Use a 10 ml pipette for this. The reaction can take 20–40 ml of the NaOH solution, and usually takes 5–10 min. The reaction is considered to be complete when the pH changes less than 0.1 units in 10 sec.

4. Immediately add crushed ice to the reaction, enough so that some of the ice remains unmelted. This cools and stabilizes the activated beads.

5. Quickly transfer the slurry to a coarse sintered glass filter on a 4-litre suction flask and, as rapidly as possible, filter to a moist cake and then wash by passing 2 litres of ice-cold distilled water through the cake, then 500 ml cold PBS. Filter to a moist cake.

6. Add enough of the moist cake to the previously prepared antibody solution such that the volume is double the original volume. This yields a final antibody concentration of 1 mg/ml in a 50% slurry of activated beads. Incubate with sufficient shaking to maintain the slurry as an even suspension for 4 h at room temperature (RT), then overnight at 4°C.

7. The next day let the beads settle and determine the OD_{260} of the remaining supernatant. Coupling efficiency is calculated as:

$$100(1 - OD_{260} \text{ final}/1.4)$$

assuming that the OD_{260} of the initial solution was 2.8. If the coupling efficiency is less than 80% there was a problem, and the coupling should be repeated with careful attention paid to maintaining the pH of the CNBr reaction above 11 and minimizing the time spent washing the beads and getting them into the antibody solution. Usually the coupling efficiency is better than 90%.

3.1.2 Preparing a crude CAM-containing sample for fractionation

The CAM of interest must be rendered soluble to be fractionated on the affinity matrix. Most CAMs are intrinsic membrane glycoproteins. Thus it is either necessary to solubilize the membranes using detergent or to isolate a soluble proteolytic fragment. This protocol assumes that you have already developed a technique for isolating plasma membranes from the cell type of interest (see Chapter 6 of this book, and *Biological Membranes: A Practical Approach*), or that you have developed and optimized digestion conditions for producing a proteolytic fragment. The sample you start with must be a clarified aqueous solution, either without detergent or with a detergent that does not interfere with the binding of antigen to antibody. Before proceeding, the sample must be clarified by centrifugation, at least 20 000 *g* for 1 h and preferably at 100 000 *g* for 1 h.

3.1.3 Running the column

The loading buffer is the same solution that was used to prepare the soluble antigen. It must not interfere with antigen–antibody binding. For elution from the affinity column a number of buffers may work. One that has worked particularly well for a number of CAMs is 50 mM diethylamine, pH 11.5. This can be supplemented with 1 mM $CaCl_2$ for isolating calcium-dependent adhesion molecules. The eluted sample is neutralized by directly collecting into 1/10 the final expected volume of 1 M Hepes, pH 7.0.

Protocol 4. Running an immunoaffinity column

Materials
- chromatography column, capable of holding 1/10 the volume of the crude extract
- loading buffer (see considerations discussed above), 40 column volumes
- elution buffer: 50 mM diethylamine, pH 11.5, with either 1 mM EDTA

Generation and characterization of function-blocking

Protocol 4. *Continued*

  or 1 mM CaCl$_2$ depending on the nature of the molecule being isolated, optional detergent
- neutralization buffer: 1 M Hepes, pH 7.0, 3 column volumes
- immunoaffinity matrix, prepared as described in *Protocol 3*
- BioBeads SM-2 (Bio-Rad): if the column is being run in detergent

Method

1. Wash the affinity resin prepared above by pouring it into a column and running through 10 column vol. of sample loading buffer, 2 column vol. of elution buffer and 10 column vol. of sample loading buffer.
2. Add to the antigen solution approximately 1/10 vol. of washed immunoaffinity resin and mix by gently swirling at 4°C for 1–2 h to bind the antigen to the beads. Stop agitation and let the beads settle for 15–30 min.
3. Pour the solution into a chromatography column with a sintered glass bottom, trying to leave the majority of the beads to the last. This effectively collects all of the immunoaffinity resin in the minimum time in the column. Rinse the last of the beads out of the flask into the column with small aliquots of sample loading buffer.
4. Wash the column with 10 column vol. of sample loading buffer. Elute with one column volume of elution buffer. If you are not monitoring the outflow then discard the first 15% of the column volume and collect one column vol. into neutralizing buffer. To save the column, wash through sample loading buffer until the pH of the flowthrough returns to that of the sample loading buffer.
5. If the sample was run in a detergent-containing solution, remove the detergent by incubating the solution with 1/4 volume of BioBeads-SM-2 for 30 min at 4°C. The BioBeads adsorb detergents and leave the protein in the aqueous phase. Dialyse the aqueous phase exhaustively against distilled water at 4°C and lyophilize. Redissolve the lyophylized material in an appropriate buffer and centrifuge to clarify. The clarified supernatant should be highly enriched in the antigen and should be checked for purity by SDS-PAGE.

3.2 Preparing the antigen for injection

For preparation of polyclonal antibodies, the specificity of the response depends critically on the purity of the antigen. Minor contaminants can cause a major immune response and result in a serum that is not specific for the antigen of choice. Although affinity purification is highly selective, a single immunoaffinity purification from a crude membrane or cell extract will usually be contaminated with low levels of other antigens. Also, some CAMs, for example the cadherins, are tightly complexed to other proteins, and these proteins must be removed prior to immunization. One way of eliminating this problem is to repeat the immunoaffinity purification. An easier approach, and one that yields a very effective immunogen, is to perform a final purification of the antigen by preparative SDS-PAGE. This enables you to excise only the material of the correct molecular weight, dissociated from virtually all contaminants that may be binding non-covalently to the antigen, and also serves to evaluate the purity of the antigen after immunoaffinity purification.

For immunization it is possible to use the excised gel band, but it is usually better to electroblot the antigen on to nitrocellulose and then inject the pulverized nitrocellulose as the antigen. This causes no long term health problems for the animal and provides a long-lasting depot of antigen. For injection and handling of animals it is essential that qualified animal-care technicians be consulted.

Protocol 5. Immunizing with protein bound to fragments of nitrocellulose

Materials

- two rabbits (or two goats if a large amount of serum is desired)
- purified protein band, transferred to nitrocellulose, 20–200 µg per injection
- complete Freund's adjuvant
- incomplete Freund's adjuvant

Method

1. Pulverize the nitrocellulose with the protein bound to it in a 50:50 mixture of complete Freund's adjuvant and PBS, using a mechanical homogenizer.
2. Inject the emulsion into two rabbits, subcutaneously between the shoulder blades and in the fold of skin at each rear leg.
3. Six weeks later, boost with the same antigen, using incomplete Freund's adjuvant, injecting at the same sites.
4. Take a blood sample seven to 10 days later and test the serum in an appropriate assay (e.g. Western blotting, ELISAs, immunocytochemical staining or direct inhibition of adhesion).
5. Repeat a cycle of boosts at 6-week intervals and blood tests 10 days later until the desired titre is obtained. At this point you can either sacrifice the animal and collect as much blood as possible, or start a regimen of collecting blood every 3–4 weeks. If the titre begins to decrease, the animal can be boosted with more antigen.

3.3 Serum treatment

The procedure described below for immunoglobulin purification will yield IgG of better than 90% purity in a form that is stable for years and can be used for any immunological techniques. The protocol uses an ammonium sulfate precipitation to generate a crude fraction that is highly enriched for IgG, then uses a DEAE–cellulose column to adsorb the major contaminants; the flowthrough is the IgG solution.

3.3.1 Preparation of serum

Blood must be collected by qualified animal-care workers. The maximum allowable volume of blood that can be withdrawn from a rabbit is 30–50 ml depending on the size of the rabbit, and this should not be done more than once every four weeks.

Protocol 6. Preparing serum from blood

1. Clot the blood by incubation at 37°C for 1–2 h.
2. Dislodge the clot from the walls of the container by running a thin glass rod around the edges and then store at 4°C for one or two days to allow the clot to contract and express the serum.
3. Decant the serum away from the clot and spin at 20 000 g for 20 min to clarify.

Protocol 7. Purification of IgG from serum

Materials
- saturated ammonium sulfate solution
- DE-52 cellulose ion-exchange resin, equilibrated in the appropriate buffer
- 17.5 mM sodium phosphate, pH 6.3 (for rabbit IgG)
- 10 mM sodium phosphate, pH 7.5 (for goat IgG)
- clarified serum

Method

1. While stirring the serum at 4°C, slowly add 0.59 vol. of saturated ammonium sulfate, continue to stir for 1 h at 4°C and leave to stand overnight at 4°C.
2. Pellet the precipitate at 12 000 g for 15 min.
3. Redissolve the pellets in half the original serum volume of distilled water.
4. Dialyse overnight against 100 vol. of 17.5 mM sodium phosphate, pH 6.3 for rabbit IgG or 10 mM sodium phosphate, pH 7.5 for goat IgG.

5. Run the crude IgG solution (up to a maximum of 25 ml) over a 40 ml column of DE-52 DEAE–cellulose that was previously equilibrated with the same phosphate buffer as in step 4 and wash through with buffer, collecting the first 200 ml of flowthrough. Dialyse extensively against distilled water and lyophilize.

The resulting material is better than 95% pure IgG, and is stable for years stored at −20°C. It can easily be dissolved in PBS and used for all immunological procedures and can be used as a starting material for preparing Fab′ fragments (see below).

3.4 Fab′ fragments

The use of intact IgGs to inhibit the function of cell surface CAM molecules can be complicated by the fact that multivalent IgG molecules can lead to capping and internalization of the antigen. In addition, the binding of a fairly large IgG molecule to an abundant cell surface molecule (like NCAM, which comprises 1% of total membrane protein) can lead to non-specific inhibition of cell–cell or membrane–membrane interactions by steric hindrance. These problems can be overcome by the use of fragments of the IgG molecule.

The IgG molecule consists of two heavy chains and two light chains, which are held together by disulfide bonds. The polypeptide chains form three functional domains, two identical antigen binding regions (Fab′)$_2$ and a constant region (Fc). IgGs can be fragmented into these functional domains by partial digestion with pepsin. Pepsin releases the two antigen binding domains still linked together [(Fab′)$_2$] as well as the Fc fragment. The (Fab′)$_2$ fragment is then reduced and alkylated to form 2 Fab′ fragments. Rabbit IgGs exhibit only a few cleavage sites for pepsin and are therefore relatively easy to handle. Mouse IgGs from different subclasses show a much wider degree of variability in cleavage sites, requiring that preliminary tests be run to assess each digestion protocol (see ref. 15 for further details).

Protocol 8. Preparation of Fab′ fragments

Materials

- lyophylized IgG
- 0.1 M Na acetate buffer, pH 4.5
- pepsin
- dialysis tubing
- 2 × PBS, pH 8.0
- β-mercaptoethanol

Protocol 8. *Continued*
- iodoacetamide
- PBS

Method

A. *Digestion of Ig with pepsin to make (Fab')$_2$*
1. Weigh out lyophylized Ig sample (must be at least 100 mg).
2. Dissolve sample in 0.1 M Na acetate buffer, pH 4.5 to 10 mg/ml.
3. Weigh out pepsin equivalent to 1/20 the weight of the Ig, dissolve in 1 ml acetate buffer and add to the Ig solution.
4. The solution should become cloudy. Mix and place in a 37°C bath overnight. The solution should mostly clear up. The initial cloudiness is due to the release of the Fc fragment, which then precipitates. Pepsin continues to digest the Fc fragment into smaller pieces which eventually dissolve and the cloudiness goes away.

B. *Reduction and alkylation of the Fab'$_2$ fragment to form Fab' fragments*
1. Centrifuge pepsin-digested sample in a table-top centrifuge at top speed for 3 min to remove any precipitate.
2. Put the sample in dialysis tubing and dialyse against 1 litre of 2 × PBS, pH 8.0, for 1 h at RT.
3. Readjust the pH of the PBS to pH 8.0 and add 6 ml of β-mercaptoethanol. Continue to dialyse for 2 h in the hood (cover well). β-mercaptoethanol reduces the disulfide bonds to two sulfhydryl groups.
4. Remove the dialysis tubing and rinse with distilled water. Place in 1 litre 2 × PBS, pH 8.0, containing 2 g of iodoacetamide. *Note:* do not get iodoacetamide on your skin! Iodoacetamide alkylates the sulfhydryl groups so disulfide bridges cannot re-form. React at room temperature for 3 h. Keep the flask covered, as iodoacetamide is light-sensitive.
5. Dialyse against 2 litres 1 × PBS, pH 7.4, 4°C, five changes, minimum. This is very important. Iodoacetamide is toxic, and if it is not completely removed from the Fab' solution it will give artefactual perturbation of tissue and organ cultures.
6. Store the contents of the dialysis tube frozen in aliquots at −20°C.

4. Testing the ability of the IgG or Fab' fragments to block function

The first step in characterizing an antibody for function blocking is to characterize its molecular specificity. The three major methods commonly used to establish the specificity of an antibody were outlined in Section 2.5.1.

4.1 Functional assays

In assessing the ability of an antibody to block CAM function, the first test is always an aggregation assay. CAMs have been operationally defined *in vitro* as molecules which mediate cell–cell or membrane–membrane aggregation. Function blocking anti-CAM antibodies or Fab fragments will thus, by definition, inhibit aggregation *in vitro*. The second step in determining the ability of an antibody to block CAM function is to use cell culture models of the *in vivo* phenomenon of interest. The two most commonly used *in vitro* model systems used to study CAM function are neuronal cultures and cultured epithelial cells or epithelial monolayers. Once the function-blocking nature of the antibody is established *in vitro*, antibodies or Fab' fragments can be introduced *in vivo* to try to eliminate CAM function in a developing system.

4.1.1 Aggregation assays
These assays are based on the ability of antibodies or Fab' fragments to inhibit cell–cell or membrane–membrane contact (see Chapter 1 for details).

4.1.2 Neuronal cell culture
In studying adhesive phenomena in neuronal culture, one of the major questions is to determine whether the cells prefer to adhere to each other or to the culture surface. In the specialized case of cultured neurones, CAMs contribute not only to adhesion among the neuronal cell bodies, but also to the growth characteristics of axons, including the initiation of a neuritic process, neurite elongation, and the fasciculation (or bundling) pattern of neurites. Both the pattern and extent of neurite growth can be influenced by CAMs, depending on which CAMs are present on the neurites and what culture substrate is used.

i. Neurite initiation and elongation
Quantitation of neurite initiation involves determining how many individual neurones have put out neurites at least one cell body diameter in length after short times (2–24 h) in culture. Neurite elongation is an extension of neurite initiation. Neurite length is estimated by making camera lucida drawings of the entire neuritic arbor and measuring neurite lengths manually or using image analysis or computerized graphic tablets. Accurate length measurements can be made only if neurones are plated at sufficiently low density to allow reliable tracing of the entire extent of the neurites and unambiguous identification of which neurones the neurites are growing from. This means that the neurones have to be completely dissociated from the ganglion, and have to be plated at very low density (200 neurones/cm^2).

Measure neurite initiation (percentage of neurones with neurites) or total neurite length in control cultures and those treated with control Fab fragments. Add function-blocking antibodies to another set of cultures and measure experimental growth.

ii. Neurite fasciculation and branching

The pattern of neurite fasciculation reflects a balance between cell–cell and cell–substrate interactions (1). If cell–cell interactions dominate, neurites will tend to grow along each other, i.e. to fasciculate, whereas if cell–substrate adhesion dominates, the neurite will tend to spend more time on the substrate and less time growing on another axon, thus the pattern of growth will be defasciculated. The degree of fasciculation of neurites is difficult to assess quantitatively. Neurite fascicles consist of numerous individual neurites, so the thickness of a bundle may be generally correlated with the number of neurites present in a bundle, but can also reflect alterations in neurite–neurite adhesion, depending, for example, on the number of different CAMs expressed by neurites. An antibody-mediated alteration in fasciculation may be missed by light microscopy and may only be detectable by electron microscopy (16). A change in branching may also be difficult to quantitate using light microscopy, because branches may remain confined in the neurite bundle. These problems have led to a tendency to analyse fasciculation and branching by purely qualitative means: neurites in one treatment group 'look' more fasciculated than in another group. The best system to use for fasciculation studies is an explant culture or a retinal strip (16, 17), in which the neurites grow out in close proximity to one another, thus having ample opportunity to interact with one another.

Treat one set of cultures with a control antibody (see below) or Fab' fragment, and the other set with the function-blocking Fab' fragment. Photographs of stained or phase views of the final pattern of neurite growth are made and compared. Alternatively, the growing neurites are photographed using time-lapse, followed by assessment of the amount of time a given growth cone spends on the substrate versus on other axons.

iii. Choice of controls

For assays of neurite growth and/or fasciculation, use of Fab fragments rather than intact IgGs is recommended because CAMs are relatively abundant surface molecules, and cross-linking of IgGs on the surfaces of cells can cause non-selective effects on neurite growth or bundling. Likewise, the selection of a control IgG or Fab fragment to add is an important experimental design issue. Pre-immune or non-immune IgG is not really an adequate control, especially if you are not using Fab fragments. The control antibody should be directed against some cell-surface epitope which is fairly abundant but not functionally important for neurite extension. Often the best control is a non-function blocking anti-CAM antibody, i.e. one that binds to the cell but does not alter the function of the CAM.

iv. Choice of growth substrate

Substrates which can have been used to study the role of CAMs in neurite growth and fasciculation *in vitro* include:

- tissue culture plastic-coated with a polycation such as poly-L-lysine or poly-DL-ornithine (18, 19)
- purified extracellular matrix components (e.g. laminin) coated on to dishes or dotted on to nitrocellulose (19, 20)
- frozen sections of various tissues (21–23)
- purified populations of non-neuronal cells, including Schwann cells (24–27), astrocytes (28, 29) and muscle cells (30, 31).

When neurones are grown in co-culture with other cell types, or on tissue sections, neurites are difficult to visualize using phase-contrast or DIC optics. In order to measure neurite length or evaluate the pattern of growth, neurones must first be immunostained with a neurone-specific marker (for example, neurofilament or synapsin I).

v. Systems that have been successfully used
Perturbation of the fasciculation patterns of neurons growing *in vivo* by anti-CAM antibodies have been carried out in two systems: motor neurones growing into the developing hindlimb of the chicken (32) and retinal ganglion cell axons growing towards the optic fissure (33). In both of these experimental paradigms, Fab' fragments were used which had been first shown to be function-blocking in both aggregation assays and in neurite fasciculation studies *in vitro*.

4.1.3 Epithelial integrity assay
Many of the CAMs, particularly the cadherins, are essential in maintaining the integrity of epithelial junctional complexes. One way to test for active inhibiting antibodies is to add Fab fragments to monolayer cultures and evaluate changes in the appearance of the cultures. Look for the cells to become more rounded and for spaces to develop between the cells. A more quantitative assay is to measure changes in transepithelial resistance in monolayers of the cell type of choice. It is important to add the antibodies to epithelial cultures from the basal side because the tight junctions at the apical side will prevent access of the antibodies to the lateral surfaces. See Chapter 11 for details of this assay.

5. Recombinant bacteriophage antibody libraries
Recent developments in molecular biology have made it possible to prepare cDNA expression bacteriophage libraries that express a huge complement of antibody specificities from mRNA prepared from spleens of immunized animals (34, 35). This essentially means that it is possible to immortalize the immune response of an animal and have the resulting library available for numerous repeated screenings and clonings by different criteria, and to

Generation and characterization of function-blocking

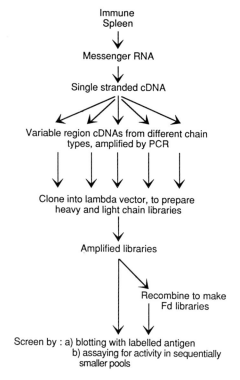

Figure 1. Schematic plan for isolating a recombinant monoclonal antibody from a bacteriophage library. Immunize a mouse and isolate spleen cells as for hybridoma preparation, but instead of fusing with myeloma cells to prepare a few immortalized hybridomas, prepare mRNA from the cells. Selectively amplify the variable regions from all of the Ig heavy and light chain mRNAs by PCR, using primers that flank the variable regions in the various chains, and insert the resulting cDNAs into appropriate bacteriophage expression vectors. Once these libraries are amplified you have what can be thought of as an immortalized immune repertoire. The single chain libraries can be recombined to give libraries that express the Fd antibody fragments, which consist of the variable regions from a heavy and light chain complexed together to form a whole antibody binding site. Screen aliquots for the desired functional activities by whatever assay is appropriate. One powerful approach is to assay pools of 10^5 to 10^6 for activity, then reassay pools of 1/10 of the complexity from active pools, and repeat this until a single clonal phage is isolated.

produce expressed antibody mixtures that are the formal equivalent of a polyclonal serum. The basic scheme for this approach is outlined in *Figure 1*.

This new technology is changing very rapidly, as new vectors and cloning strategies are being developed. There is one commercially available kit (Stratacyte) that provides the reagents and instructions for preparing such antibody expression libraries in lambda bacteriophage. Stratacyte has a fairly rigorous licensing agreement that requires purchasers of the kit not to divulge the details of their procedures. However, the kit is complete and has an excellent manual, so it may well prove to be the simplest approach for a project that just requires one or two libraries, if you do not want to devote your complete efforts to learning all the techniques from scratch.

References

1. Jessell, T. M. (1988). *Neuron,* **1**, 3.
2. Watanabe, M., Frelinger, A. L., and Rutishauser, U. (1986). *J. Cell Biol.,* **103**, 1721.
3. Takeichi, M. (1987). *Trends Genet.,* **3**, 213.
4. Harlow, E. and Lane, D. (ed.) (1988). *Antibodies: A Laboratory Manual.* Cold Spring Harbor Laboratory, Cold Spring Harbor, NY.
5. Freund, J. (1956). *Adv. Tuberc. Res.,* **7**, 130.
6. Hawkes, R. (1986). *Methods in Enzymology,* Vol. 121 (ed. J. J. Langone and H. Van Vunakis) pp. 484–91. Academic Press, Orlando, Florida.
7. Engvall, E. and Perlmann, P. (1972). *J. Immunol.,* **109**, 129.
8. Towbin, H., Staehelin, T., and Gordon, J. (1979). *Proc. Natl Acad. Sci. USA,* **76**, 4350.
9. Lane, D. P. and Lane, E. B. (1981). *J. Immunol. Methods,* **47**, 303.
10. Tsu, T. T. and Herzenberg, L. A. (1980). In *Selected Methods in Cellular Immunology* (ed. B. B. Mishell and S. M. Shiigi), pp. 373–97. W. H. Freeman, San Francisco.
11. Richman, D. D., Cleveland, P. H., Oxman, M. N., and Johnson, K. M. (1982). *J. Immunol.,* **128**, 2300.
12. Kohler, G. and Milstein, C. (1976). *Eur. J. Immunol.,* **6**, 511.
13. Oi, V. T. and Herzenberg, L. A. (1980). In *Selected Methods in Cellular Immunology* (ed. B. B. Mishell and S. M. Shiigi), pp. 351–72. W. H. Freeman, San Francisco.
14. Suzuki, S., Sano, K., and Tanihara, H. (1991). *Cell Reg.,* **2**, 261.
15. Mage, M. G. (1980). *Methods in Enzymology,* Vol. 70 (ed. H. van Vanukis and J. L. Langone), pp. 142–50. Academic Press, New York.
16. Rutishauser, U., Gall, C., and Edelman, G. M. (1978). *J. Cell Biol.,* **79**, 382.
17. Bonhoeffer, F. and Huf, J. (1982). *EMBO J.,* **1**, 427.
18. Yavin, E. and Yavin, Z. (1974). *J. Cell Biol.,* **62**, 540.
19. Ure, D. and Acheson, A. (1992). In *Neuromethods.* Vol. 21: *Cell Cultures* (ed. A. A. Boulton, G. B. Baker, and W. Walz). (In press.)
20. Edgar, D., Timpl, R., and Thoenen, H. (1984). *EMBO J.,* **3**, 1463.
21. Carbonetto, S., Evans, D., and Cochard, P. (1987). *J. Neurosci.,* **7**, 610.
22. Covault, J., Cunningham, J. M., and Sanes, J. R. (1987). *J. Cell Biol.,* **105**, 2479.
23. Sandrock, A. W. and Matthew, W. D. (1987). *Proc. Natl Acad. Sci. USA,* **84**, 6934.
24. Bixby, J. L., Lilien, J., and Reichardt, L. F. (1988). *J. Cell Biol.,* **107**, 353.
25. Kleitman, N., Wood, P., Johnson, M. I., and Bunge, R. P. (1988). *J. Neurosci.,* **8**, 653.
26. Seilheimer, B. and Schachner, M. (1988). *J. Cell Biol.,* **107**, 341.
27. Letourneau, P. C., Shattuck, T. A., Roche, F. K., Takeichi, M., and Lemmon, V. (1990). *Devel. Biol.,* **130**, 430.

28. Tomaselli, K. J., Neugebauer, K. M., Bixby, J. L., Lilien, J., and Reichardt, L. F. (1988). *Neuron,* **1,** 33.
29. Smith, G. M., Rutishauser, U., Silver, J., and Miller, R. H. (1990). *Devel. Biol.,* **138,** 377.
30. Bixby, J. L. and Reichardt, L. F. (1987). *Devel. Biol.,* **119,** 363.
31. Bixby, J. L., Pratt, R. S., Lilien, J., and Reichardt, L. F. (1987). *Proc. Natl Acad. Sci. USA,* **84,** 2555.
32. Landmesser, L., Dahm, L., Schultz, K., and Rutishauser, U. (1988). *Devel. Biol.,* **130,** 645.
33. Thanos, S., Bonhoeffer, F., and Rutishauser, U. (1984). *Proc. Natl Acad. Sci. USA,* **81,** 1906.
34. Huse, W. D., Sastry, L., Iverson, S. A., Kang, A. S., Alting-Mees, M., Burton, D. R., Benkovic, S. J., and Lerner, R. A. (1989). *Science,* **246,** 1275.
35. Kang, A. S., Barbas, C. F., Janda, K. D., Benkovic, S. J., and Lerner, R. A. (1991). *Proc. Natl Acad. Sci. USA,* **88,** 4363.

3

Expression cloning: transient expression and rescue from mammalian cells

ALEJANDRO ARUFFO

1. Introduction

Cloning strategies based on transient expression in mammalian cells have been used effectively to isolate cDNAs encoding secreted, cell surface, and intracellular proteins. The first application of transient expression in mammalian cells in cloning was the isolation of cDNAs encoding the lymphokine GM-CSF (granulocyte macrophage colony stimulating factor) (1, 2). This was followed by the use of transient expression in mammalian cells in conjunction with antibody panning (3) to isolate cDNAs encoding cell surface proteins (4, 5). Transient expression in mammalian cells has also been used to isolate cell surface receptors based on the ability of the transfected cells to bind to either fluorescein-conjugated (6) or radiolabelled (7) ligands or ligand immobilized on a solid phase (8, 9) or ligand expressed on a cell surface (10). Recently, transient expression in mammalian cells has also been used to isolate cDNAs encoding intracellular proteins (11, 12).

This chapter describes the use of transient expression in the simian cell line COS (13) in conjunction with an immunologic screening on antibody coated dishes to isolate cDNAs encoding the antigens against which the antibodies are directed. The transfection and selection protocols can also be applied to isolating clones that produce binding to other cells or extracellular matrix proteins that are immobilized on a culture dish.

2. Overview of the cloning procedure

Expression cloning of cDNAs encoding cell-surface proteins using COS cells and panning is carried out by sequentially applying four rounds of 'transient expression and rescue'. Each round of transient expression and rescue contains the following steps (*Figure 1*): *Step 1*, transfecting COS cells with cDNA subcloned in the appropriate expression vector; *Step 2*, treating the transfected

Expression cloning

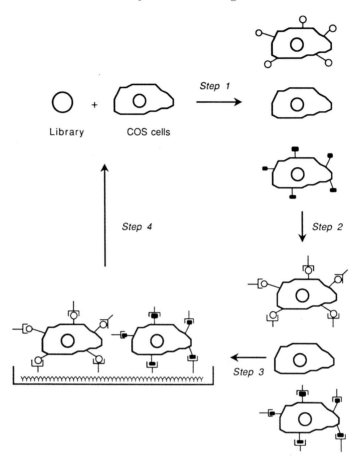

Figure 1. Transient expression and rescue. Isolation of cDNA clones encoding cell surface protein using transient expression and rescue. A cDNA library prepared in the mammalian expression vector (or equivalent) is transfected into COS cells (*Step 1*). Forty-eight hours post-transfection the cells are incubated with antibodies directed against the cell surface proteins of interest (*Step 2*). The excess antibodies are washed from the transfected cells and the positive cells culled from the bulk of the transfected cells by panning on an antibody coated dish (*Step 3*). The panning dish is washed and plasmid DNA is recovered from the adherent cells (*Step 4*). This DNA is amplified in *E. coli* and reintroduced into COS cells to initiate another round of transient expression in COS cells and immunological selection.

cells (48 h post-transfection) with antibodies which recognize the proteins of interest; *Step 3*, culling the immunoreactive COS cells expressing the protein of interest from the bulk of the transfected cells by panning (3) on antibody coated dishes; and *Step 4*, recovering the plasmid DNA containing the cDNAs encoding the protein of interest from the panned cells using the method of Hirt (14). This plasmid DNA is subsequently amplified in

Escherichia coli (*E. coli*) and used to initiate another round of transient expression and rescue (step 1).

Two different transfection protocols are used with this procedure. The first round of transient expression and rescue is initiated by transfecting COS cells using the diethylaminoethyl (DEAE)–dextran method (4, 5). This transfection procedure allows the efficient transfection of COS cells (40–70% of the cells can be transfected with this procedure) ensuring complete cDNA library representation in the transfected cells. However, this transfection procedure delivers multiple cDNA types into each cell compromising the ability of panning to enrich for the cDNA of interest. For this reason, transfection by spheroplast fusion (4, 5) is used in all the subsequent rounds. Spheroplast fusion is not very efficient (1–5% of the cells can be transfected by this procedure); however, it does ensure that each transfected COS cell receives a single cDNA type, allowing the enrichment of the cDNA encoding the protein of interest by panning (3).

The rescue of the cDNA encoding the protein of interest is carried out by first incubating the transfected cells with either monoclonal or polyclonal antibodies against the protein or proteins of interest 48 h post-transfection. After antibody incubation the excess antibody is washed away and positive cells are isolated from the bulk of the transfected cells by panning (3) on antibody coated dishes. Plasmid DNA containing the clone of interest can be recovered from the panned cells using the method of Hirt, amplified in *E. coli*, and used to initiate another cycle of transient expression and rescue.

Although this protocol is designed with COS cells in mind there are times when the antibodies to be used cross-react with proteins expressed by COS cells. In this case COS cells should not be used since the panning steps will provide no enrichment. However, it is possible to substitute COS cells with a murine equivalent such as WOP cells (15, 16) which would not cross-react with the antibodies.

3. Vectors

A number of different vectors have been developed for transient expression in mammalian cells and COS cells in particular. These have recently been reviewed by Kaufman (17). The most important elements within these vectors is the presence of a strong enhancer/promoter that will give high-level expression in the transfected cells of the subcloned cDNA and the presence of elements that allow the vector to replicate to a high copy number in the transfected cells. This high copy number enhances the level of protein expression in the transfected cells and permits the rescue of the plasmid from the COS cells, using the method of Hirt (14).

Here the expression vector CDM8 is described (*Figure 2*) (16). This vector can be used with either COS (13) cells or WOP (15) cells. It contains: (*i*) both an SV40 and polyoma viral origin of replication which allows the vector to

Expression cloning

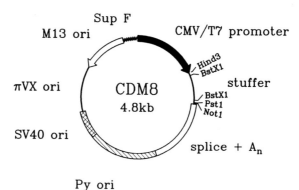

Figure 2. CDM8 expression vector. The CDM8 vector (16) contains (i) an SV40 and polyoma viral origins of replication (SV40 ori and Py ori); (ii) an *E. coli* and M13 origins of replication (πVX ori and M13 ori); (iii) a *Sup F* gene for bacterial selection; (iv) a cytomegalovirus/T7 RNA polymerase enhancer/promoter element (CMV/T7); (v) two BstX1 subcloning sites; and (vi) an SV40 virus derived intron and poly(A) addition signal (splice + A_n).

replicate in cells which express SV40 virus (COS cells) or polyoma virus large T antigen (WOP or MOP cells) (15, 18); (*ii*) a prokaryotic origin of replication; (*iii*) an M13 origin of replication which permits the preparation of single-stranded DNA; (*iv*) a *SupF* gene which allows selection of the plasmid in a non-suppressing host containing the plasmid, p3, which contains amber mutated ampicillin and tetracycline drug-resistance elements; (*v*) the human cytomegalovirus enhancer promoter fused to a T7 RNA polymerase promoter, to drive the expression of the subcloned cDNA and to prepare RNA transcripts *in vitro*; and (*vi*) an SV40 virus derived intron and a polyadenylation (poly(A)) signal sequence.

4. Methods included in this chapter

This chapter will cover methods for: (*i*) the propagation of the expression vector CDM8 (Section 5); (*ii*) the preparation of mRNA (Section 6); (*iii*) the synthesis of cDNA and the preparation of a size fractionated cDNA library in the expression vector CDM8 (Section 7); and (*iv*) the screening of cDNA libraries by transient expression and rescue in COS cells (Section 8).

5. Propagation of SupF plasmids

The p3 plasmid is derived from RP1, is 57 kb in length, and is stably maintained as a single-copy episome. Selection for tet resistance alone is almost as good as selection for amp+tet resistance. However, spontaneous appearance of chromosomal suppressor tRNA mutations presents an unavoidable background (frequency at 10^{-9}) in this system. Colonies arising from spontaneous suppressor mutations are usually bigger than colonies arising from plasmid transformation. Bacterial strains containing p3 include MC1061/p3, FG2, and XS127 (4, 5, 16).

Suppressor plasmids are selected in LB medium containing amp at 12.5 μg/ml and tet at 7.5 μg/ml. For large plasmid preparations we use M9 casamino acids medium containing glycerol (0.8%) as a carbon source, and grow the bacteria to saturation.

6. mRNA preparation

One of the most important obstacles in the way of successful expression cloning is the preparation of cDNA libraries containing full-length cDNA copies of the mRNAs present in the cell expressing the protein of interest. In many cases, the failure to obtain a clone using transient expression cloning systems is the result of a poor library that does not contain a full-length cDNA copy of the gene of interest. The first step in preparing a high-quality cDNA library is preparing high-quality mRNA. Here two methods for the preparation of mRNA are described. The first is a two-step approach which involves the isolation of total RNA by a modification of the method of Chirgwin *et al.* (19) followed by isolation of poly(A)$^+$ using an oligo-dT column. This variation on the method of Chirgwin *et al.* (19) raises the capacity and speed of the conventional guanidinium SCN/CsCl gradient method. The second is a single step method for the isolation of mRNA from a proteinase K/SDS cell lysate.

Protocol 1. Guanidinium thiocyanate/LiCl protocol for total RNA isolation

Materials
- GuSCN/LiCl solution: for each millilitre of mix desired, dissolve 0.5 g GuSCN in 0.58 ml of 25% LiCl (stock filtered through 0.45-micron filter) and add 20 μl of β-mercaptoethanol. The solution should not be kept for more than one week at room temperature (RT)
- CsCl solution: 5.7 M CsCl solution (RNase-free; 1.26 g CsCl added to every millilitre of 10 mM EDTA)
- RNase-free water
- 3 M NaOAc
- ethanol
- SW55 or SW28 (Beckman) rotors and tubes, or equivalent

Methods
1. Harvest cells and concentrate by spinning.

Expression cloning

Protocol 1. Continued

2. Disperse pellet on walls by flicking tubes and add 1 ml of GuSCN solution to up to 5×10^7 cells.
3. Shear by polytron or with a needle and syringe until non-viscous.
4. For small-scale preparations ($<10^8$ cells) layer up to 3.5 ml of sheared mix on 1.5 ml of CsCl solution in an RNase-free SW55 polyallomer tube. If needed, overlay the GuSCN solution with RNase-free water to fill each tube to the top and spin in SW55 rotor at 50 000 r.p.m. for 2 h.
5. For large-scale preparations, layer 25 ml of sheared mix on 12 ml CsCl solution in RNase-free SW28 tubes. If needed, overlay the GuSCN solution with RNase-free water to fill the tube to the top and spin in SW28 rotor at 24 000 r.p.m. for 8 h.
6. Aspirate contents carefully with a sterile Pasteur pipette connected to vacuum flask. Once past the CsCl interface, scratch a band around the tube with the pipette tip to prevent the layer on the wall of the tube from creeping down. Aspirate remaining CsCl solution.
7. Take pellets up in water, don't try to redissolve. Add 1/10 vol. NaOAc, 3 vol. EtOH, spin. Resuspend pellet in water, at 70°C if necessary. Adjust concentration to 1 mg/ml and freeze.

Protocol 2. Poly(A)$^+$ RNA preparation from total RNA

Materials
- loading buffer: 0.5 M LiCl, 10 mM Tris pH 7.5, 1 mM EDTA, 1% SDS
- middle wash buffer: 0.15 M LiCl, 10 mM Tris pH 7.5, 1 mM EDTA, 1% SDS
- 0.1 M NaOH
- 10 M LiCl
- RNase-free polypropylene disposable column (wash with 5 M NaOH and then rinse with RNase-free water)
- 2 mM EDTA, 0.1% SDS
- RNase-free water
- 3 M NaOAc
- ethanol
- oligo-dT cellulose
- SW55 rotor and tubes, or equivalent

Methods
1. For each milligram total RNA use approx. 0.3 ml (final packed bed) oligo-dT cellulose. Prepare oligo-dT cellulose by resuspending approx.

0.5 ml of dry powder in 1 ml of 0.1 M NaOH and transferring it into the column, or by percolating 0.1 M NaOH through a previously used column (columns can be reused many times). Wash with several column volumes of RNase-free water, until pH is neutral. Rinse with 2–3 ml of loading buffer, then remove column bed to sterile 15-ml tube using 4–6 ml of loading buffer.

2. Heat total RNA to 70°C for 2–3 min.
3. Add LiCl from an RNase-free stock to 0.5 M, and combine with oligo-dT cellulose in 15-ml tube.
4. Vortex or agitate for 10 min. Pour into column.
5. Wash with 3 ml loading buffer.
6. Wash with 3 ml of middle wash buffer.
7. Elute mRNA directly into an SW55 tube with 1.5 ml of 2 mM EDTA, 0.1% SDS, discarding the first two or three drops.
8. Precipitate eluted mRNA by adding 1/10 vol. 3 M NaOAc and filling tube with EtOH.
9. Mix, chill for 30 min at $-20°C$, and spin at 50 000 r.p.m. at 5°C for 30 min.
10. Pour off EtOH and air-dry pellet.
11. Resuspend mRNA pellet in 50–100 µl of RNase-free water.
12. Analyse the quality of the mRNA by melting 5 µl of poly(A) RNA at 70°C in Mops/EDTA/formaldehyde and run on an RNase-free 1% agarose gel.

Protocol 3. Alternative mRNA preparation

Materials
- proteinase K solution: proteinase K 200 µg/ml (pre-digested for 2 h at 50°C from a 20 mg/ml stock)
- 0.5% SDS, 0.1 M NaCl, 20 mM Tris pH 7.5, and 1 mM EDTA
- oligo-dT cellulose
- disposable polypropylene columns
- high-salt buffer (0.5 M NaCl, 1 mM EDTA, 0.1% SDS)
- low-salt buffer (1 mM EDTA, 0.1% SDS)
- RNase-free water
- 5 M NaCl stock solution

Methods
1. Homogenize cells (10^9 cells) in 50 ml of proteinase K solution with a Polytron.
2. Incubate at 37°C for 1 h.

Expression cloning

Protocol 3. *Continued*

3. Add NaCl to 0.5 M from 5 M stock.
4. Add oligo-dT in high-salt buffer and mix with end-over-end mixer for 2 h at RT.
5. Spin and remove supernatant.
6. Wash twice with 5 ml of high-salt buffer.
7. Pour into disposable polypropylene column and wash two more times with 5 ml of high-salt buffer.
8. Elute with three column volumes of low-salt buffer.
9. Heat eluate at 65°C for 5 min.
10. Add NaCl to 0.5 M.
11. Pour over a new oligo-dT cellulose column equilibrated with high-salt buffer.
12. Reheat the flow-through at 65°C for 3 min and reapply to the column.
13. Wash the column twice with 5 ml of high-salt buffer.
14. Elute poly(A) RNA with 3 column vol. of low-salt buffer.
15. Precipitate by adding three cumulative volumes of ethanol.
16. Pour off ethanol and air-dry pellet and resuspend in 50–100 µl of RNase-free water.

---

7. cDNA synthesis and library preparation

The cDNA synthesis procedure presented in *Protocol 3* is a variation on the method of Gubler and Hoffman (20). Although we use oligo-dT to prime our cDNA synthesis in this protocol, random primers could also be used to initiate cDNA synthesis. After cDNA synthesis, BstX1 adaptors are added to the cDNA, to allow subcloning into the expression vector. This is followed by size fractionation on a KOAc velocity gradient. Typically, gradient fractions containing cDNAs of 1 kb or more are pulled and used to prepare the library.

The cDNA is subcloned into the vector using the non-self complementary BstX1 adaptor system (5) (*Figure 3*). Two identical BstX1 cleavage sites in inverted orientation were placed in the CDM8 vector and separated by a short stuffer fragment. Digestion of the vector with BstX1 yields a vector that is capable of accepting cDNAs that have BstX1 adaptors but not to itself, allowing the efficient ligation of cDNA to the vector.

The efficient preparation of a cDNA library requires that the ligation of the cDNA to the vector needs to be calibrated by using a constant amount of cDNA and variable amounts of vector to identify the ratio of cDNA and

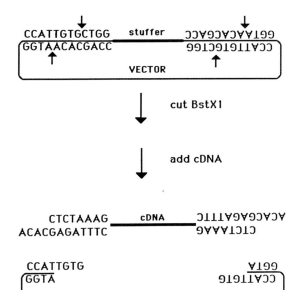

Figure 3. BstX1 subcloning adaptor system. The adaptors used for subcloning cDNA into the expression vector are based on the BstX1 restriction enzyme recognition site (CCAN$_5$/NTGG). Two identical BstX1 sites (CCATTGTGCTGG) were introduced head to head in the expression vector and separated by a stuffer fragment. Digestion of the vector with BstX1 and removal of the stuffer fragment results in a vector that cannot ligate to itself but can be ligated to cDNAs which have been ligated to adaptors and has overhangs compatible to the overhangs left after BstX1 digestion of the vector.

vector to be used in the large scale ligation for the preparation of the library. The goal should be to prepare a library of about 1×10^6 clones.

Protocol 4. cDNA synthesis, adaptor ligation, and size fractionation

Materials
- RT1 buffer: 0.25 M Tris pH 8.8 (8.2 at 42°C), 0.25 M KCl, 30 mM MgCl$_2$
- RT2 buffer: 0.1 M Tris pH 7.5, 25 mM MgCl$_2$, 0.5 M KCl, 0.25 mg/ml BSA, 50 mM DTT

Expression cloning

Protocol 4. *Continued*

- 10 × low-salt: 60 mM Tris pH 7.5, 60 mM $MgCl_2$, 50 mM NaCl, 2.5 mg/ml BSA 70 mM β-ME
- 10 × ligation additions: 1 mM ATP, 20 mM DTT, 1 mg/ml BSA 10 mM spermidine
- RNase-free water
- RNase inhibitor (Boehringer, 36 U/μl)
- oligo-dT (Collaborative Research)
- 20 mM dNTP solution (US Biochemical)
- 1 M dithiothreitol
- reverse transcriptase (RT-XL, Lifesciences, 24 U/μl)
- DNA polymerase I (Boehringer, 5 U/μl)
- RNase H (BRL 2 U/μl)
- T4 DNA ligase (New England Biolabs, 400 U/μl or Boehringer, 1 Weiss unit/μl)
- 0.5 M EDTA, pH 8
- 0.8 μg/μl 12-mer adaptor oligonucleotide, kinased
- 0.8 μg/μl 8-mer adaptor oligonucleotide, kinased
- prepare a 20% KOAc, 2 mM EDTA, 1 μg/ml EthBr solution and a 5% KOAc, 2 mM EDTA, 1 μg/ml EthBr solution: the KOAc stock solution should be filtered through a 0.45 μ (or 0.22 μ) filter.
- linear polyacrylamide, used as an inert carrier for ethanol precipitation, can be prepared by polymerizing acrylamide in the absence of bisacrylamide (but with the usual amounts of TEMED, ammonium persulphate, etc.) to make a 5% stock polymer solution.

Methods

A. cDNA synthesis: First strand

1. In RNase-free microfuge tube add 4 μg of mRNA and heat to approx. 100°C for 30 sec to denature.
2. Quench on ice, and adjust volume to 70 μl with RNase-free water.
3. Add 20 μl of RT1 buffer, 2 μl of RNase inhibitor, 1 μl of 5 μg/μl of oligo-dT, 2.5 μl of 20 mM dNTPs, 1 μl of 1 M DTT and 4 μl of reverse transcriptase.
4. Incubate at 42°C 40 min.
5. Heat inactivate the reaction at 70°C 10 min.

B. cDNA synthesis: Second strand

1. To heat inactivated first-strand synthesis add 320 μl of RNase-free water, 80 μl of RT2 buffer, 5 μl of DNA polymerase I, 2 μl of RNase H.

2. Incubate at 15°C for 1 h and 22°C for 1 h.
3. Add 10 μl of 0.5 M EDTA pH 8.0.
4. Phenol extract with an equal volume of phenol and EtOH precipitate by adding NaCl to 0.5 M, linear polyacrylamide (carrier) to 20 μg/ml, and filling tube with EtOH.
5. Spin 2–3 min in microfuge.
6. Pour off ethanol, air-dry pellet, and resuspend in 240 μl of TE.

C. *Adding adaptors to cDNA*
1. Add 30 μl of 10 × low salt buffer, 30 μl of 10 × ligation additions, 3 μl (2.4 μg) of kinased 12-mer adaptor, 2 μl (1.6 μg) of kinased 8-mer adaptor, and 1 μl of T4 DNA ligase.
2. Incubate at 15°C overnight.

D. *Size fractionation*
1. Add 2.6 ml of 20% KOAc solution to back chamber of a small gradient maker. Remove air bubble from tube connecting the two chambers by allowing solution to flow into the front chamber and then tilt back. Close passage between chambers, and add 2.5 ml of the 5% solution to the front chamber. If there is liquid in the tubing from a previous run, allow the 5% solution to run just to the end of the tubing, and then return to chamber. Place the apparatus on a stirplate, set the stir bar moving as fast as possible, open the stopcock connecting the two chambers and then open the front stopcock. Fill a polyallomer SW55 tube from the bottom with the KOAc solution. Overlay the gradient with 100 μl of cDNA solution.
2. Prepare a balance tube in the same way as the sample tube.
3. Spin the gradient for 3 h at 50 000 r.p.m. at 22°C.
4. Collect fractions from the gradient, pierce the SW55 tube with a butterfly infusion set (with the luer hub clipped off) close to the bottom of the tube and collect three 0.5 ml fractions and then six 0.25 ml fractions into microfuge tubes (about 22 and 11 drops respectively).
5. EtOH precipitate the fractions by adding linear polyacrylamide to 50 μg/ml and filling the tube to the top with EtOH.
6. After cooling tubes spin in microfuge for 3 min, pour off ethanol and rinse pellets with 70% ethanol.
7. Resuspend each 0.25 ml fraction in 10 μl of TE.
8. To analyse, run 1 μl on a 1% agarose minigel to determine the contents of each fraction.
9. Pool the first three fractions, and those of the last six which contain no material smaller than 1 kb.

Expression cloning

Protocol 5. Preparation of vector for cloning and library preparation

Materials
- Bs X1 (New England Biolabs) and recommended buffer
- 50 °C water bath
- gradient maker, 5 ml capacity
- 20% and 5% KOAc solutions, prepared as described above (*Protocol 4*)
- Beckman SW55 rotor or equivalent
- polyallomer SW55 tubes or equivalent
- 1-ml syringe and 20-gauge needle
- linear polyacrylamide (see *Protocol 4*)

Methods
1. Cut 20 µg of vector in a 200 µl reaction with 100 units of Bs X1 at 50 °C overnight in a well thermostatted water bath (i.e. circulating water bath).
2. Prepare 2 KOAc 5–20% gradients in SW55 tubes as described above (*Protocol 4*, D) for size fractionating cDNA.
3. Add 100 µl of the digested vector to each tube and run for 3 h, 50 000 r.p.m. at 22 °C.
4. Examine the tube under 300 nm UV light. The desired band will have migrated ⅔ of the length of the tube and remove the band with a 1-ml syringe and 20-gauge needle.
5. Add linear polyacrylamide and precipitate the plasmid by adding 3 vol. of EtOH.
6. Resuspend in 50 µl of TE.

Prior to setting up the full-scale ligation between vector and cDNA it is important to calibrate the vector/cDNA ligations

Protocol 6. Subcloning DNA into the expression vector

Materials
- 10 × low-salt buffer (see *Protocol 4*)
- 10 × ligation additions (see *Protocol 4*)
- electroporation competent cells
- electroporator
- 24 × 24 cm plastic dishes (Nunc)

Methods

1. Set up trial ligations using a constant amount of cDNA and increasing amounts of vector, as in *Protocol 4*. Typically, use 25 ng of vector DNA for each ligation. After ligation transfect bacteria with aliquots of each ligation and determine the titres by plating on selective plates (see Section 5).
2. On the basis of these trial ligations set up large-scale ligation. Usually the entire cDNA preparation requires 1–2 μg of cut vector.
3. The library is plated on ten 24 × 24 cm dishes at 10^5 colonies/dish.
4. Twenty-four hours after plating the library is harvested and episomal DNA prepared and purified on a CsCl gradient.

8. Transient expression screening of the cDNA library

The procedures given in this section include methods for: (*i*) COS cell transfection by the DEAE–dextran method (21), which is used in the initial round of transfection (*Protocol 7*); (*ii*) panning to immunoselect COS cells which carry the cDNA encoding the protein of interest (*Protocols 8* and *9*); (*iii*) recovering the plasmid of interest from the cells which have bound to the antibody-coated panning dish by the method of Hirt (*Protocol 10*); and (*iv*) COS cell transfection by the spheroplast fusion method (22), which is used in rounds 2–4 of transfection (*Protocol 11*).

Typically a library of more than 1×10^6 clones will be screened by transfecting COS cells with plasmid DNA using the DEAE–dextran method (the initial cycle should be carried out with ten, 50% confluent dishes of COS cells, see below). The DEAE–dextran method is a high-efficiency transfection procedure that will ensure good library representation in the transfected cells. Forty-eight hours post-transfection, the COS cells expressing the protein of interest are immunoselected from the bulk of the transfected cells by the method of panning. First, the cells are lifted from the dish, washed, and incubated with antibodies recognizing the proteins of interest. These cells are subsequently washed to remove the excess antibody and placed on a Petri dish coated with anti-antibody antibodies. This procedure allows for the simultaneous isolation of cDNAs encoding different proteins. The panning dishes are then washed to remove non-adherent cells and plasmid DNA is prepared from the adherent cells. This DNA is amplified in *E. coli* and used to initiate a second round of transient expression and rescue.

All subsequent rounds of transient expression and rescue are initiated by transfecting the COS cells by the spheroplast fusion method. Although inefficient, this method ensures that a single plasmid type gets delivered into each transfected COS cell, allowing efficient enrichment for the plasmid of interest by panning. If antibodies against a single protein are being used in the panning step it is only necessary to carry out two additional rounds of transient expression and rescue. If, on the other hand, the panning step is being carried out with antibodies against different proteins an additional fourth round of transient expression and rescue is required in which the last panning step is carried out with antibodies against each of the proteins separately.

8.1 DEAE–dextran transfection

COS cells to be transfected are split the night before, to give 50% confluence the day of transfection. One simple way to achieve this is to resuspend the trypsinized cells from a confluent 10-cm dish in 10 ml of medium, and distribute 2 ml into each 10-cm dish to be transfected. The higher the degree of confluence, the greater the ability of the cells to withstand the transfection conditions; however, expression following transfection at high density is typically inferior to that at lower density.

Cells are transfected in DMEM or IMDM containing 10% heat-inactivated NuSerum (Collaborative Research). NuSerum allows the cells to withstand the transfection conditions for greater lengths of time, resulting in improved net expression. If calf or fetal bovine serum is used, a thick precipitate forms in a substantial fraction of transfections, which results in severely eroded cell viability and expression. Presumably the low protein concentration in the NuSerum obviates this.

Protocol 7. DEAE–dextran transfection

Materials
- 100 mm tissue-culture dishes seeded with COS cells
- 1 × DMEM + 10% NuSerum
- 1 × DMEM + 10% CS
- 1 × PBS + 10% DMSO
- 25 × DEAE–dextran/chloroquine solution in PBS (10 mg/ml DEAE–dextran and 2.5 mM chloroquine)

Methods
1. For each 10-cm dish to be transfected use 5 ml of DMEM or IMDM/NuSerum.
2. Add 5 µg of DNA and mix.
3. Add 0.2 ml of DEAE–dextran/chloroquine 25 × stock solution and mix.
4. After 4 h at 37 °C the DEAE-containing medium is removed by aspiration,

and 2 ml of 10% DMSO in PBS added. It is generally important to check the cells after about 3 h of exposure to the DEAE transfection mix, as their health can decline precipitously. This is particularly true of chloroquine transfections, and it is usually better to shorten the transfection than to allow too many cells to die.

5. After 2 min or longer at room temperature (timing is not essential), the PBS/DMSO is removed and replaced with fresh medium (10 ml of DMEM/100 CS).

6. The next day, the transfected COS cells should be trypsinized and replated on to a new 10-cm dish.

8.2 Panning

This method is based on the procedure described by Wysocki and Sato (3). This step can also be carried out with a cell sorter. However, panning is less expensive (plastic dishes vs. cell sorter) and more efficient, especially since COS cells are quite big and must be run through the cell sorter at slow flow rates.

Protocol 8. Preparing antibody-coated dishes

Materials
- 60-mm bacteriological plates, Falcon 1007 or equivalent, or 10-cm dishes such as Fisher 8-757-12
- sheep anti-mouse affinity purified antibody from Cooper BioMedical (Cappell) or equivalent
- 0.15 M NaCl
- 1 mg/ml BSA in PBS

Methods
1. Dilute anti-mouse antibody to 10 mg/ml in 50 mM Tris–HCl, pH 9.5.
2. Add 3 ml per 6-cm dish, or 10 ml per 10-cm dish.
3. Incubate at RT for 1.5 h (this antibody solution can be used on two more dishes).
4. Wash plates three times with 0.15 M NaCl.
5. Incubate with 3 ml 1 mg/ml BSA in PBS overnight at RT.
6. Aspirate the PBS/BSA blocking solution and store plates in −20°C freezer.

Expression cloning

Protocol 9. Separating cells by panning

Materials
- 60-mm panning dishes
- PBS/0.5 mM EDTA/0.02% azide
- PBS/0.5 mM EDTA/0.02% azide/5% FBS
- PBS/0.5 mM EDTA/0.02% azide/2% Ficoll
- PBS/5% FBS
- 100-micron nylon mesh (Tetko)
- mouse antibody against the protein that is being cloned[a]

Methods
1. Aspirate medium from dish with the transfected COS cells, add 2 ml PBS/0.5 mM EDTA/0.02% azide and incubate dishes at 37°C for 30 min to detach cells from dish.
2. Triturate cells vigorously with short Pasteur pipette, and collect cells from each dish in a centrifuge tube.
3. Spin 4°C, 200 g, 5 min.
4. Resuspend cells in 0.5–1.0 ml PBS/EDTA/azide/5% FBS and add antibodies.
5. Incubate >30 min on ice.
6. Add an equal volume of PBS/EDTA/azide, layer carefully on 3 ml PBS/EDTA/azide/Ficoll, spin 4 min 200 g.
7. Aspirate supernatant in one smooth movement.
8. Take up cells in 0.5 ml PBS/EDTA/azide and add aliquots to antibody-coated dishes, containing 3 ml PBS/EDTA/azide/5% FBS, by pipetting through the nylon mesh.
9. Add cells from at most two 60-mm dishes to one 60-mm antibody-coated plate.
10. Let sit at RT 1–3 h.
11. Remove excess cells not adhering to dish by gentle washing with PBS/5% serum or with medium. Two or three washes of 3 ml are usually sufficient.

[a] Purified antibodies should be at a concentration of 1 μg/ml, and ascites used at a final dilution of 1/1000. These concentrations may have to be modified; generally concentrations that are used for FACS analysis are good for this procedure.

Once the cells have been selected it is necessary to isolate the episomal DNA and transform it into bacteria for the next round of cloning.

Protocol 10. Preparing a Hirt supernatant

Materials
- lysis solution, 0.6% SDS, 10 mM EDTA
- 5 M NaCl
- linear polyacrylamide (see *Protocol 4*)
- ethanol
- 3 M NaOAc
- TE, 10 mM Tris pH 7.5, 1 mM EDTA
- electroporation competent cells
- electroporator

Methods
1. Add 0.4 ml 0.6% SDS, 10 mM EDTA to panned plate.
2. Let sit 20 min (can be as little as 1 min if there are practically no cells on the plate).
3. Pipette viscous mixture into microfuge tube.
4. Add 0.1 ml 5 M NaCl, mix, put on ice at least 5 h. Keeping the mixture as cold as possible seems to improve the quality of the Hirt.
5. Spin 4 min, remove supernatant carefully, phenol extract (twice if the first interface is not clean), add 10 μg linear polyacrylamide (or other carrier), fill tube to top with EtOH, precipitate, and resuspend in 0.1 ml. Add 10 μl 3 M NaOAc and 330 μl EtOH, reprecipitate and resuspend in 0.1 ml of TE.
6. Transform into MC1061/p3 using electrocompetent cells.

8.3 Spheroplast fusion

This procedure is based on the method described by Sandri-Goldrin *et al.* (22). A set of six fusions requires 100 ml of cells in broth.

Protocol 11. Transfection by protoplast fusion

Materials
- 50% PEG/50% DMEM solution, adjust pH to 7 with 7.5% sodium bicarbonate (Baker or Kodak PEG 1000 or 1450)
- 100 mg/ml spectinomycin or 150 mg/ml chloramphenicol
- 20% sucrose, 50 mM Tris pH 8.0
- 5 mg/ml lysozyme (Sigma) freshly dissolved in 0.25 M Tris–HCl, pH 8

Expression cloning

Protocol 11. *Continued*
- 0.25 M Tris pH 8.0
- 50 mM Tris pH 8.0
- 1 × DMEM, 10% sucrose, 10 mM $MgCl_2$
- 1 × DMEM
- 1 × DMEM, 10% CS, 15 μg/ml gentamicin sulfate
- Sorvall RT60000B table-top centrifuge with a H1000B rotor

Methods
1. Grow bacterial cells containing amplifiable plasmid to $OD_{600} = 0.5$ in LB.
2. Add spectinomycin to 100 μg/ml (or chloramphenicol to 150 μg/ml).
3. Continue incubation at 37°C with shaking for 10–16 h. (Cells begin to lyse with prolonged incubation in spectinomycin or chloramphenicol medium.)
4. Spin down 100 ml of culture (JA14/GSA rotor, 250-ml bottle) 5 min at 10 000 r.p.m.
5. Drain well, resuspend pellet in bottle with 5 ml cold 20% sucrose, 50 mM Tris–HCl pH 8.0.
6. Add 1 ml of 5 mg/ml lysozyme.
7. Incubate on ice 5 min.
8. Add 2 ml cold 0.25 M EDTA pH 8.0, incubate 5 min on ice.
9. Add 2 ml 50 mM Tris pH 8, incubate 5 min at 37°C (water bath).
10. Place on ice, check percentage conversion to spheroplasts by microscopy.
11. In flow hood, slowly add 20 ml of cold DMEM/10% sucrose/10 mM $MgCl_2$ (dropwise, about 2 drops per second).
12. Remove media from COS cells plated the day before in 6-cm dishes (50% confluent).
13. Add 5 ml of spheroplast suspension to each dish.
14. Place dishes on top of tube carriers in swinging bucket centrifuge. Up to six dishes can be prepared at once. (Dishes can be stacked on top of each other, but three in a stack is not advisable as the spheroplast layer on the top dish is often torn or detached after centrifugation.) Spin at 100 g (setting 5.7) 10 min. (Force is calculated on the basis of the radius of the bottom plate.)
15. Aspirate fluid from dishes carefully.
16. Pipette 1.5–2 ml 50% PEG/50% DMEM (no serum) into the centre of the dish. If necessary, sweep the pipette tip around to ensure that the PEG spreads evenly and radially across the whole dish.

17. After PEG has been added to the last dish, prop all of the dishes up (on their lids) so that the PEG solution collects at the bottom.
18. Aspirate the PEG. (The thin layer of PEG that remains on the cells is sufficient to promote fusion; the layer remaining is easier to wash off, and better cell viability can be obtained, than if the bulk of the PEG is left behind.)
19. After 90 to 120 sec (PEG 1000) or 120 to 150 sec (PEG 1450) of contact with the PEG solution, pipette 1.5 ml of DMEM (NuSerum serum) into the centre of the dish. The PEG layer will be swept radially by the DMEM.
20. Tilt the dishes and aspirate.
21. Repeat the DMEM wash.
22. Add 3 ml of DME/10% serum containing 15 µg/ml gentamicin sulfate.
23. Incubate 4–6 h in incubator.
24. Remove media and remaining bacterial suspension, add more media and incubate 2–3 days.

Although most protocols call for extensive washing of the cell layer to remove PEG, this washing tends to remove many of the cells without any substantial benefit. If the cells are allowed to sit in the second DMEM wash for a few minutes, most of the spheroplast layer will come up spontaneously. The remaining spheroplasts come off within 4–6 h in the complete medium at 37°C.

The resulting transformed cells are grown up and then panned for expression (see *Figure 1* for overall scheme).

9. Conclusion

The procedures described in this chapter have been used to isolate cDNAs encoding cell surface proteins using antibodies against these proteins as probes. Complementary DNA libraries prepared in expression vectors like CDM8 can also be screened with radiolabelled ligands (7), or antibodies (12), or functionally (1, 2, 8–11) to isolate cDNAs encoding proteins of interest. Together these screening procedures offer a large number of alternatives to isolate cDNAs encoding secreted (1, 2), cell surface (4–10), and intracellular proteins (11, 12) (for a review, see ref. 23) from expression libraries. However, if the protein of interest is part of a heterocomplex and the presence of two or more members of the complex are required for expression of the protein, the cDNA encoding the protein will be refractory to cloning using transient expression systems like the one described in this chapter.

Acknowledgements

I am in debt to Brian Seed; the protocols presented in this chapter are largely based on a set of protocols that were prepared by Brian. The proteinase K/SDS protocol for the isolation of mRNA was provided by Herb Lin. I also thank Ana Wieman and Mary West for help in preparing this manuscript.

References

1. Lee, F., Yokota, T., Otsuka, T., Gemmell, L., Larson, N., Luh, J., Arai, K. I., Rennick, D. (1985). *Proc. Natl Acad. Sci. USA*, **82**, 4360.
2. Wong, G. G., Witek, J. S., Temple, P. A., Wilkens, K. M., Leary, A. C., Luxenberg, D. P., Jones, S. S., Brown, E. L., Kay, R. M., Orr, E. C., Shoemaker, C., Golde, D. W., Kaufman, R. J., Hewick, R. M., Wang, E. A., and Clarck, S. C. (1985). *Science*, **228**, 810.
3. Wysocki, L. J. and Sato, V. L. (1978). *Proc. Natl Acad. Sci. USA*, **75**, 2844.
4. Seed, B. and Aruffo, A. (1987). *Proc. Natl Acad. Sci. USA*, **84**, 3365.
5. Aruffo, A. and Seed, B. (1987). *Proc. Natl Acad. Sci. USA*, **84**, 8573.
6. Yamasaki, K., Taga, T., Hirata, Y., Yawata, H., Kawanishi, Y., Seed, B., Taniguchi, T., Hirano, R., and Kishimoto, T. (1988). *Science*, **241**, 825.
7. Sims, J. E., March, C. J., Cosman, D., Widmer, M. B., MacDonald, H. R., McMahan, C. J., Grubin, C. E., Wignall, J. M., Jackson, J. L., Call, S. M., Friend, D., Alpert, A. R., Gillis, S., Urdal, D. L., and Dower, S. K. (1988). *Science*, **241**, 585.
8. Stengelin, S., Stamenkovic, I., and Seed, B. (1988). *EMBO J.*, **7**, 1053.
9. Staunton, D. E., Dustin, M. L., and Springer, T. A. (1989). *Nature (Lond.)*, **339**, 61.
10. Osborn, L., Hession, C., Tizard, R., Vassallo, C., Luhowskyj, S., Chi-Rosso, G., and Lobb, R. (1989). *Cell*, **59**, 1203.
11. Tsai, S. F., Martin, D. I., Zon, L. I., D'Andrea, A. D., Wong, G. G., Orkin, S. H. (1989). *Nature (Lond.)*, **339**, 446.
12. Metzelaar, M. J., Wijngaard, P. L., Peters, P. J., Sixma, J. J., Nieuwenhuis, H. K., Clevers, H. C. (1991). *J. Biol. Chem.*, **266**, 3239.
13. Gluzman, Y. (1981). *Cell*, **23**, 175.
14. Hirt, B. (1967). *J. Mol. Biol.*, **26**, 365.
15. Dailey, L. and Basilico, C. (1985). *J. Virol.*, **54**, 739.
16. Seed, B. (1987). *Nature (Lond.)*, **329**, 840.
17. Kaufman, R. J. (1990). In *Methods in Enzymology*, Vol. 185 (ed. D. V. Goeddel), pp. 537–66. Academic Press, Orlando, Florida.
18. Muller, W. J., Naujokas, M. A., and Hassell, J. A. (1984). *Mol. Cell. Biol.*, **4**, 2406.
19. Chirgwin, J. M., Przybyla, A. E., MacDonald, R. J., and Rutter, W. J. (1979). *Biochemistry*, **18**, 5294.
20. Gubler, U. and Hoffman, B. J. (1983). *Gene*, **25**, 263.
21. Sussman, D. J. and Milman, G. (1984). *Mol. Cell. Biol.*, **4**, 1641.
22. Sandri-Goldrin, R. M., Goldrin, A. L., Lewine, M., and Glorioso, J. C. (1981). *Mol. Cell. Biol.*, **1**, 743.
23. Aruffo, A. (1991). *Current Opinion in Biotechnology*, **2**, 735.

4

Functional analysis of uvomorulin by site-directed mutagenesis and gene transfer into eukaryotic cells

JORG STAPPERT and ROLF KEMLER

1. Introduction

The generation and expression of suitable mutants of a gene is a powerful technique for altering the primary structure of the coded protein to study structure–function relationships. New methods for the isolation and cloning of genes and, above all, the chemical synthesis of DNA allow genes to be altered specifically by introducing deletions or insertions or by replacing individual bases in the DNA. In this way virtually any change desired may be made to the protein primary structure involving the 20 amino acids. The biological activity of these mutants can then be tested *in vitro* and *in vivo*. The availability of full-length cDNAs coding for cell adhesion molecules (CAMs) has opened new possibilities for detailed studies on the molecular mechanisms of cell adhesion.

In this chapter we will describe *in vitro* methods that we have used to generate mutants of the mouse cell-adhesion molecule uvomorulin and techniques necessary to transfect these mutant genes into various eukaryotic cell lines. These techniques are generally applicable to any other protein for which a cloned cDNA is available.

2. Background

A proper understanding of the mutants we have made requires an introduction into the structure and function of mouse uvomorulin, as briefly outlined below.

Uvomorulin is an integral membrane glycoprotein of 120 kd involved in the adhesion of embryonal and epithelial cells and belongs to the group of Ca^{2+}-dependent CAMs (1, 2). Members of this group express their adhesive properties only in the presence of Ca^{2+}. Ca^{2+} protects these proteins from proteolytic degradation (3). Comparison of the primary structures of several members of the group revealed a gene family termed cadherins. The extracellular part of uvomorulin, which mediates the selective cell adhesiveness, is composed largely of three domains with internal homology (4). Each domain contains two putative Ca^{2+} binding motifs. These motifs are well conserved and are located at analogous positions in all cadherins, suggesting that they might interact with Ca^{2+}. To test this possibility mutant uvomorulins with amino acid substitutions in the putative Ca^{2+} binding 'motif B' of the amino-terminal domain of uvomorulin were created, using complementary oligonucleotides (5). This method will be described in Section 3.1.

The cytoplasmic region is the most conserved part of all Ca^{2+}-dependent CAMs, suggesting a common functional role. This view found experimental support when we showed that the cytoplasmic domain of uvomorulin is associated with three structurally independent proteins with molecular masses of 102, 88, and 80 kd, which are named α, β, and γ catenin respectively (6). Catenins seem to be present nearly ubiquitously in different cell types.

To study the association of catenins with the cytoplasmic region of the uvomorulin we generated mutant uvomorulin polypeptides, with single amino acid substitutions within the cytoplasmic region of uvomorulin to delete putative phosphorylation sites which might interact with the catenins. This was done using the polymerase chain reaction (PCR) as described in Section 3.2.

The generation of mutations within a coding gene can be accomplished by the 'recombinant PCR' technique, as described in Section 3.3. We used this technique to delete the complete propeptide sequence of uvomorulin in order to characterize the function of the leader sequence. All Ca^{2+}-dependent CAMs are synthesized as precursor polypeptides, but in contrast to the extended homology between the mature proteins the respective precursor segments show no structural similarity. Each precursor segment varies in length, the sequence comparison reveals a rather weak homology of 28–40% at the nucleotide level. For uvomorulin the precursor segment is composed of 129 amino acids, which are cleaved off intracellularly (7).

Another variation of the basic PCR technique is presented in Section 3.4. This method was used to generate mutant uvomorulin polypeptides with deletions at the 3'-end of the cytoplasmic part of the protein in order to define exactly the region with which catenins interact.

In the second part of this chapter we will describe the three techniques (electroporation, calcium phosphate precipitation, and lipotransfection) for the transfection of DNA into eukaryotic cells.

We are aware of the fact that it is impossible to describe all methods currently available for site-directed mutagenesis *in vitro* (for reviews see refs 8–10). The methods we have chosen for this chapter are just a few examples of how to create various mutations of various kinds. These techniques are, however, general and powerful and can be the basis for most directed mutagenesis experiments.

3. General aspects

Practical approaches concerning the construction of the various mutants mentioned above will be given exclusively for the mutational step itself. Commonly used techniques like plasmid preparation, ligation, transformation, or the elution of fragments were done according to Maniatis *et al.* (11). Sequence analysis of the mutants was done on double-stranded DNA, using T7 Sequenase Kit (US Biochemical) according to the manufacturer's description.

The synthetic oligonucleotides used for the construction of the mutants should be of the highest purity and, depending on the enzymes used to subclone the mutated DNA fragments, it may be necessary to quantitatively phosphorylate them at the 5'-terminus. Such oligonucleotides may be synthesized by a number of chemical methods and are also available from several commercial suppliers. The oligonucleotides we used were synthesized on an Applied Biosystems Model 394 DNA Synthesizer and then purified using a C-18 reversed-phase high-pressure liquid chromatography (HPLC) column.

3.1 Site-directed mutagenesis by complementary oligonucleotides

A rapid method to generate a mutation within a given cDNA sequence is outlined in *Figure 1*. The two complementary strands for an oligonucleotide carrying two types of mutation are synthesized: the desired mutation to substitute, delete or insert base pairs and a 'silent', tagging mutation which creates or deletes a restriction site that can be used to identify the mutant clones. The strands are constructed to produce appropriate restriction sites at their ends, and the synthetic DNA is then used to replace the corresponding fragment in the target clone by ligating it into appropriately cut target DNA.

To create a silent mutation, it is necessary to make single base-pair substitution that does not affect the amino acid sequence of the protein. By computer analysis it is fast and easy to check the DNA sequences for possible sites to create silent mutations. After replacement of the wild-type sequence by the mutated oligonucleotides, positive clones can easily be identified by restriction digestion due to the presence of the silent marker.

In general, this method should be applied to mutations which are closely flanked by restriction sites or if several mutations are to be set within a short piece of DNA. Otherwise, the cost and difficulty of synthesizing large synthetic oligonucleotides becomes prohibitive.

The annealing and ligation conditions we used in this context are listed in *Protocol 1*. Plasmid DNA was prepared using the alkaline lysis method,

Functional analysis of uvomorulin

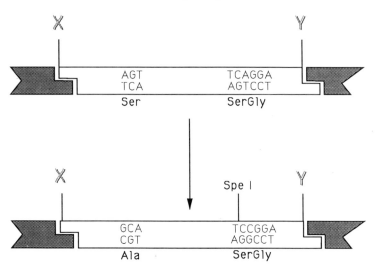

Figure 1. Introduction of mutations using synthetic oligonucleotides. Two kinds of mutations are introduced into a short stretch of synthetic DNA that corresponds to a region flanked by two restriction sites X and Y in the target DNA. In the example shown the first mutation leads to the amino acid substitution of Ser to Ala, whereas the second mutation affects only the nucleotide sequence. With this 'silent mutation' a new restriction site sequence for SpeI is generated which can be used as 'selection marker' to facilitate finding positive clones.

digested with the appropriate restriction enzymes and then purified. Using two different restriction enzymes generates non-compatible ends and makes it unnecessary to dephosphorylate the vector or to phosphorylate the oligonucleotides, thus avoiding the formation of oligonucleotide concatemers. But when using identical protruding termini the oligonucleotides still have to be phosphorylated (see *Protocol 1*).

Protocol 1. Introduction of mutations by insertion of synthetic oligonucleotides

Materials
- oligonucleotides designed for the appropriate alterations
- ligation reagents (see Maniatis ref. 11)
- competent *Escherichia coli*

Methods
1. Determine the concentration of the oligonucleotides by measuring the OD at 260 nm.[a]
2. Mix an equal number of moles of the two oligonucleotides in a test-tube.

3. Heat to 70°C for 10 min; cool slowly down to room temperature, within 30 min.
4. Ligate 0.04 pmol of the digested plasmid DNA with a 10 to 20-fold excess of the annealed oligonucleotides carrying compatible ends in a small volume (10–20 µl), 4 h at 16°C or alternatively at 4°C overnight.
5. Use 3–5 µl to transform competent *E. coli* cells.
6. Check several clones for the presence of the selection marker by restriction digestion.
7. If the oligonucleotides have to be phosphorylated, set up the following reaction:

    200 pmol oligonucleotide
    3 µl 1 M Tris–HCl pH 8.0
    1. 5 µl 0.2 M $MgCl_2$
    0.1 M DTT
    1 mM ATP pH 7.0
    4.5 U T4 polynucleotide kinase
    Add double distilled water to 30 µl
    Incubate at 37°C for 45 min
    Heat to 65°C for 10 min

[a] A_{260} of 1.0 = 30 µg/ml

3.2 Single point mutation by PCR

The polymerase chain reaction (PCR) offers a wide spectrum of applications to alter DNA sequences (12, 13). Deletions, insertions, and point mutations can be constructed *in vitro* by oligonucleotide-mediated site directed mutagenesis procedures in which appropriate oligonucleotides containing the mutation are used to prime DNA synthesis. The method is diagrammed in *Figure 2* and outlined in detail in *Protocol 2*.

Oligonucleotide primers with the desired sequence changes are synthesized and used to amplify the target region from a cloned template. The amplified fragment, which now contains the changes that were introduced into the primers, is then inserted into the appropriate position in the target DNA by subcloning. This approach makes it possible to introduce various mutations near a restriction site, and the choice of the second restriction site for subcloning can be varied. This has the advantage of allowing one to mutate DNA sequences even in cases where one restriction site is far from the mutation site. Again the introduction of a selective marker will facilitate finding positive clones.

Unfortunately, the approach to the selection of efficient and specific primers for PCR remains somewhat empirical. There is no set of rules that

Functional analysis of uvomorulin

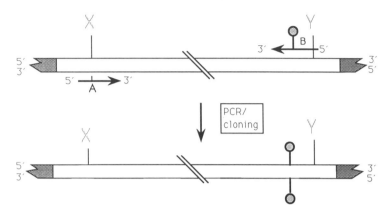

Figure 2. Scheme for the construction of mutants by PCR. Two oligonucleotide primers are represented by the arrows adjacent to their annealing sites in the target sequence. Both primers A and B span appropriate restriction site sequences. To generate amino acid substitutions, oligonucleotide B contains one or several mismatches, which will be incorporated into the amplification product. Furthermore, insertions and deletions can also be introduced into the amplified product with the appropriate primers.

will ensure the synthesis of an effective primer pair, but the following guidelines will help in their design:

(a) If possible, select primers with a random base distribution and a GC content of 50–60%.
(b) Avoid sequences with significant secondary structures, particularly at the 3'-end of the primer.
(c) To ensure a strong binding of the 3'-end of the primer, try to terminate with two or three G/C pairs
(d) Primers should have an overall length of 15–30 bases.
(e) A few extra bases (usually two) are added 5' to the restriction site to ensure that the efficiency of restriction enzyme cleavage is maintained.
(f) A total of 6–8 bases of exact homology to the template should be included on each end of the oligonucleotides to ensure proper hybridization of the primer.

For all PCR methods we describe in this chapter double-stranded plasmid DNA was used, prepared according to the alkaline method already mentioned.

Protocol 2. Generation of mutations in DNA fragments synthesized by PCR

Materials
- $10 \times$ *Taq* buffer is: 100 mM Tris–HCl pH 8.4, 500 mM KCl, 20 mM $MgCl_2$ and 1 mg/ml gelatine

- 1 × *Taq* dilution/storage buffer is: 20 mM Tris–HCl pH 8.0, 100 mM KCl, 0.1 mM EDTA, 1 mM DTT, 0.5% NP-40, 50% glycerol and 0.5% Tween-20

Methods

1. In a 0.5-ml test-tube set up 100 μl of the following reaction mixture:
   7×10^{-3} pmol double-stranded plasmid DNA
   10 μl 10 × *Taq* buffer
   10 μl 3' primer (10 μM)
   10 μl 5' primer (10 μM)
   10 μl dNTPs (each at 1000 μM in this stock; neutralize dNTPs to pH 7.4)
   Double-distilled water to 100 μl

   Overlay with 60 μl mineral oil to prevent evaporation.

2. Heat to 95°C for 3 min.
3. Add 0.5 U of *Taq* polymerase (diluted in *Taq* storage/dilution buffer).
4. Set up the folowing PCR conditions:
   Anneal at 55°C for 1 min
   Extend at 72°C for 1.5 min
   Denature at 95°C for 1 min
5. Repeat the last three steps 30 times.
6. Extract the completed PCR with 1 vol. of chloroform to remove the mineral oil.
7. Check 5–10 μl of the PCR product by gel electrophoresis.
8. Mix 20–30 μl of the PCR product with 5 μl of the appropriate 10 × restriction buffer, 1–2 units of the restriction enzymes, and double-distilled water to 50 μl.
9. Incubate at the optimal temperature for the enzymes 1–2 h.
10. Inactivate the enzymes by heating to 70°C if they are heat labile, or by extraction with an equal volume of buffered phenol if they are not.
11. To remove dNTPs carried over from the PCR which may inhibit ligation, and, since DNA needs to be concentrated before ligation, we recommend purifying the DNA by either isopropanol precipitation in the presence of potassium acetate (0.3 M final concentration pH = 4.8) or by adsorption to and elution from Quiagen columns (Diagen).
12. Ligation is carried out according to Maniatis *et al.* (11).

3.3 Generating mutations within a gene

The method described in this section is a variation of the basic PCR technique and can be used to create specific site substitutions, insertions, and deletions at any position in a gene.

Scharf et al. (14) first showed that it was quite simple to introduce restriction site sequences into DNA fragments produced by PCR merely by attaching these sequences to the 5'-end of oligonucleotides used as primers. As strands initiated by these 'add-on' primers are themselves copied, the added restriction site sequence becomes 'fixed' into the growing population of PCR product fragments. These add-on sequences can also be used to recombine DNA fragments at any desired junction, as shown in *Figure 3A* and *Figure 3B*, where two PCR products from different sections of the same template are made such that the resultant fragments overlap in sequence. These products can be mixed, denatured and allowed to reanneal. One of the heteroduplex forms consists of DNA strands that overlap at their 3'-ends. Only these heteroduplices can be extended by the *Taq* polymerase and therefore be amplified in the second PCR. Combined with primer-introduced sequence modification, this process of PCR fragment joining allows a sequence alteration at any position in the fragment, not just at the ends as described in previous methods (15, 16, 17).

The primer used for this purpose should have at least 15 bases of target sequence complementarity 3' to any add-on sequence. Add-on sequences that provide overlap between PCR products should have a length of at least 12–15 bases. In order to favour production of the combined PCR product, it is necessary to remove the overlapping or 'inside' primers from the primary PCR products and then add the 'outside' primer. Therefore, products of the first PCR reaction should be purified by gel electrophoresis.

Figure 3. (**A**) Site-specific mutagenesis and combining PCR fragments that overlap in sequence. The two middle or 'inside' primers A_{in} and B_{in} are overlapping primers, each carrying complementary 5 'add on' sequences that can be used to create a deletion relative to the target sequence by combining the overlapping PCR products. PCR A and PCR B are performed separately. The products are separated from excess primers and mixed, denatured, and allowed to reanneal. Some of the molecules recombine as shown through the overlap made by the middle primers. One of the heteroduplex forms consists of DNA strands that overlap at their 3'-ends. DNA chain extension of these recombinants with 3' recessed ends leads to a molecule that can be amplified with the original outside primers A_{out} and B_{out}. (**B**) Agarose gel electrophoresis of PCR amplification products. To delete the whole leader sequence of uvomorulin, which has an overall length of 137 amino acids, two overlapping inside primers were used. Each of the two primers has 16 or 17 bases of target sequence complementarity with a GC content of 63–65%. The 5' add-on sequences that provide overlap between the PCR products were 15 to 16 bases in length. The amount of GC was 60%. The outer primers were 16 to 18 bases in length with 50 to 60% GC. PCR A and PCR B were done separately, yielding amplification products of 255 bp and 388 bp in length (*lanes 2* and *3*). The products were separated from excess primers and mixed, denatured and allowed to reanneal. Amplification of the recombined PCR products by the original outside primers resulted in a DNA fragment of 643 bp (*lane 4*). Lane 1 = 123 bp ladder from Gibco-BRL.

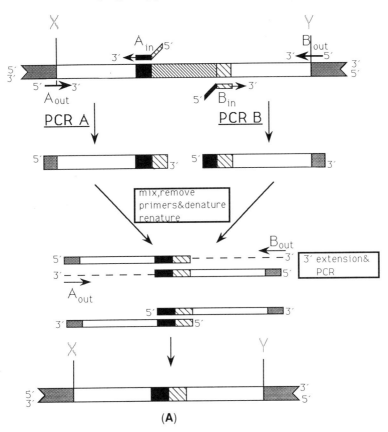

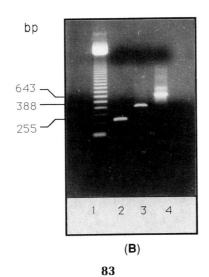

(A)

(B)

Protocol 3. Introduction of internal mutations by PCR

Oligonucleotides designed as described above.

Method
1. For each PCR follow the procedure given in *Protocol 2*, steps 1–7.
2. Purify 30 µl of each PCR product by gel electrophoresis.
3. After DNA-precipitation dissolve the DNA in 10 µl double-distilled water.
4. Remix 4 µl of each PCR product. (It is important to have the same molarities of each PCR product.)
5. Set up a second PCR with the two 'outside' primers.
6. Follow the description given for the PCR conditions in *Protocol 2*, steps 1–12.

3.4 Deletion mutants at the 3'-end of a gene

This method has also been used in our group, to make exact 3' deletion mutants of uvomorulin, as shown in *Figure 4A*. A restriction site sequence as well as a stop codon to terminate translation were added to the 5'-end of several primers. For the PCR a second primer located further upstream and containing a restriction site sequence was used. PCR conditions as described in *Protocol 2* were used to amplify the DNA fragment. The fragments were then subcloned back into the target DNA using the two restriction sites. The result is shown in *Figure 4B*.

These few examples of site-directed mutagenesis done by PCR show the wide spectrum of applications offered by this method. Nevertheless, changes introduced into the sequence of the PCR products due to nucleotide misincorporation can create problems. In an initial study the error rate was calculated to be 2×10^{-4} nucleotide/cycle (18). This is in reasonable agreement with the report of a 10^{-4} nucleotide error rate determined for the *in vitro* replication of a β-galactosidase template by the *Taq* polymerase (19). Although different reaction conditions may influence the error rate, this estimated error rate does not pose a problem for most applications. None the less we recommend sequencing the final clones derived from PCR.

4. Introduction of recombinant vectors into mammalian cells

To analyse mutated genes it is necessary to introduce the eukaryotic DNA into cultured mammalian cells. Many transfection methods have been developed for this purpose. The most commonly used techniques are still DNA calcium phosphate coprecipitation, electroporation, or gene transfer

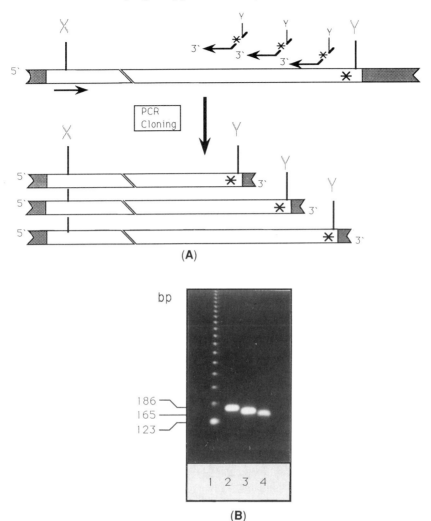

Figure 4. (A) Scheme for the construction of 3' mutants of uvomorulin. One universal primer spanning the restriction site sequence X and several 3' oligonucleotides, each with a stop codon and a restriction site sequence Y added to their 5'-end were used to generate various 3' deletion mutants. The PCR products with the deletions are then cloned into the target DNA using the restriction sites. (B) Small 3' deletions of uvomorulin, 8-, 15-, and 22- amino-acids respectively, were done by PCR. A universal 5' oligonucleotide primer spanning a ClaI site and three 3' mutagenic primers, each with a 5' add-on sequence containing a stop codon (*) and a HindIII site, were used to amplify DNA fragments of 186 bp (lane 2), 165 bp (lane 3), and 144 bp (lane 4) in length. Subcloning was done as described in Table two. The DNA marker used was the 123 bp ladder (GIBCO-BRL) (lane 1).

using liposomes. This section will briefly describe the basic calcium phosphate precipitation and the electroporation used for stable gene transfer of non-viral vectors. We will give only a brief introduction of a new transfection technique based on lipopolyamine-coated DNA. Since selection must be based on a phenotypic change which can be scored following the transfer of cellular DNA into a recipient cell, it is advantageous to use a plasmid marker in a co-transfection with the mutated cDNA (20). Thus by first placing the transfected cells under selection with a dominant vector any background from non-transfected cells can be eliminated. This selection can be removed if necessary, and the second screening can proceed, based on the phenotypic change to be scored. Several selection markers have been described (21, 22). A widely used dominant selectable marker is resistance to neomycin, a bacterial antibiotic which interferes with prokaryotic ribosomes. While mammalian cells are not affected by neomycin, an analogue to these drugs will effect eukaryotic ribosomes. This analogue G418 is available through Gibco as Geneticin. The neoR gene codes for a phosphotransferase which inactivates the G418. Therefore, cells which express this bacterial resistance marker under eukaryotic control, can survive in selective media. This selection can be used on almost any cell type. In co-transfection experiments we usually use the vector pSVtkneoß in which the neoR gene is under the control of the thymidine kinase promoter (23). To circumvent co-transfection with two constructs, vectors such as pL31N are used, where the neoR gene is under the control of the polyoma enhancer/tk-promoter while mutated cDNA is under the control of the CMV-IE-promoter. A number of eukaryotic vectors carrying the neoR gene are now commercially available.

4.1 Transfection of coprecipitates of calcium phosphate and DNA

This is the most widely used method (24). It can be applied very successfully to most cells in monolayer and some cells in suspension. Although the mechanism remains obscure, it is believed that the transfected DNA enters the cytoplasm of the cell by endocytosis and is transferred to the nucleus. The protocol for calcium phosphate co-transfection given in *Protocol 4* is the basic transfection method without any additional treatments and should be used to transfect adherent cells (11).

Protocol 4. Transfection of adherent cells with calcium phosphate

Materials
- 10 × Hepes-buffered saline (HBS): 8.18% NaCl(w/v), 5.94% Hepes (w/v), 0.2% Na_2HPO_4 (w/v). The solution can be stored in aliquots at 4°C. Use the 10 × HBS solution to prepare 2 × HBS solution and adjust the pH to 7.12 with 1 N NaOH. Sterilize the solution through a nitrocellulose

membrane. Care should be taken in preparing this buffer since the pH is very critical

- 2 M $CaCl_2$, sterilized and stored at 4°C

Method

1. On the day before transfection replate cells to be used at a density of 10^4/cm^2 (5×10^5 cells per 9-cm plate).
2. Incubate the cultures for 20–24 h at 37°C in a humidified incubator in an atmosphere of 5–10% CO_2.
3. Replace the culture medium with fresh medium containing 10% FCS.
4. For each 60-mm monolayer of cells to be transfected prepare the calcium phosphate precipitate as follows:
   (a) warm up the stock solutions to room temperature.
   (b) set up 0.5 ml of the calcium phosphate precipitate for a 60-mm dish with 5 ml of medium; 1 ml of the precipitate is normally added to 10 ml of medium in a 90-mm dish.
   (c) In tube A place a solution containing 10–15 μg of DNA together with 31 μl of 2 M $CaCl_2$ and bring the final vol. to 0.25 ml with water.
   (d) In tube B add 0.25 ml of 2 × HBS.[a]
   (e) To make the precipitate add the DNA solution to the HBS dropwise, with gentle mixing. The precipitate will form immediately.
   (f) Transfer the calcium phosphate DNA suspension into the medium above the cell monolayer. Rock the dish gently to mix the medium.
   (g) At that time additional treatments (with reagents such as chloroquine, glycerol or sodium butyrate) can be made.
5. After incubating the cells for 16–24 h remove the medium and wash the monolayer once with phosphate-buffered saline (PBS) and add 5–10 ml pre-warmed complete growth medium.
6. Return the cells to the incubator for 16–24 h before replating the cells in the appropriate selective medium for the isolation of stable transformants.
7. Change medium every 2–3 days for 2–3 weeks to remove the debris of dead cells and to allow colonies of resistant clones to grow.
8. Individual colonies may be cloned and propagated for assay.

Published procedures differ widely in the manner and rate of mixing of ingredients, because one has to avoid the rapid formation of coarse precipitates that would result in a decreased efficiency of transformation. Other factors such as the purity of the DNA and the exact pH of the HBS-buffer affect the size of the precipitate.

To obtain the highest transformation efficiencies, plasmid DNA should be purified by equilibrium centrifugation in CsCl–ethidium bromide density gradients.

Geneticin (Gibco-BRL) is added to the culture medium at a range of 200 μg/ml–1 mg/ml. It is advisable to establish for each cell line the amount of Geneticin necessary to kill non-transfected cells in one week, while allowing resistant colonies to form by 10–14 days. Geneticin can be kept as a stock solution at 100 mg/ml and stored at −20°C.

4.2 DNA transfection by electroporation

The use of electric fields to introduce DNA into cells in culture has been termed electroporation (25, 26). In principle, this method utilizes short electric pulses of a certain field strength, which alter the permeability of membranes in such a way that DNA molecules can enter the cell. Electroporation frequently gives results in cell lines that are refractory to other techniques. However, considerable work may be required to define optimal conditions for the particular cell line under study. The efficiency of transfection by electroporation is influenced by a number of factors (11). For example field strength and pulse length appropriate for maximum recovery of transfectants must be determined for each cell type. Good starting conditions are pulse time constants of 5–20 msec obtained by suspending the cells (10^6–10^7/ml) in 0.8 ml PBS or growth medium (+-serum) in a 0.4 cm cuvette and using a 250, 500, or 960 μF capacitor. At these pulse lengths the optimal field for animal cells is usually between 500 and 1000 V/cm. Cell death of from 20–80% will accompany effective electropermeabilization. A number of different electroporation instruments are commercially available, and the manufacturers provide detailed protocols for their use.

4.3 Lipofection of lipopolyamine-coated plasmids

This method is based on a method published by J. P. Behr et al. (27). In principle they have developed a transfection method based on compacted cationic lipid-coated plasmid DNA. To this end they synthesized various lipospermines whose headgroups interact strongly with DNA and coat it with a cationic lipid layer, thus promoting the binding to the cell membranes. This technique has the advantage of being simple, efficient, and free of any detectable toxic side-effects.

An equivalent reagent, called Transfectam, is now commercially available from IBF. Transfectam is also a lipopolyamine molecule to which a spermine group is fixed in the terminal position by covalent attachment. Therefore, it is a strongly positively charged molecule and has a high affinity for DNA. Only 5 μg DNA are recommended for transfection, and the transfection time can be decreased to 30 min. However, the efficiency always depends on the host/plasmid system and on the preparation of the plasmid. Optimization is needed to obtain the best results, but the efficiency we obtained was much higher than with the calcium phosphate co-precipitation method.

References

1. Takeichi, M. (1990). *Annu. Rev. Biochem.*, **59**, 237.
2. Kemler, R., Ozawa, M., and Ringwald, M. (1989). *Curr. Opin. Cell Biol.*, **1**, 892.
3. Hyafil, F., Babinet, C., and Jacob, F. (1981). *Cell*, **26**, 447.
4. Ringwald, M., Schuh, R., Vestweber, D., Eistetter, H., Lottspeich, F., Engel, J. et al. (1987). *EMBO J.*, **6**, 3647.
5. Ozawa, M., Engel, J., and Kemler, R. (1990). *Cell*, **63**, 1033.
6. Ozawa, M., Baribault, H., and Kemler, R. (1989). *EMBO J.*, **8**, 1711.
7. Ozawa, M. and Kemler, R. (1990). *J. Cell Biol.*, **111**, 1645.
8. Botstein, D. and Shortle, D. (1985). *Science*, **229**, 1193.
9. Craik, C. S. (1985). *BioTechniques*, **3**, 12.
10. Smith, M. (1985). *Annu. Rev. Genet.*, **19**, 423.
11. Sambrook, J., Fritsch, E. F., and Maniatis, T. (1989). *Molecular Cloning: A Laboratory Manual* (2nd edn). Cold Spring Harbor Laboratory Press.
12. White, T. J., Arnheim, N., and Erlich, H. A. (1989). *TIG*, **5**, 185.
13. Oste, C. (1988). *BioTechniques*, **6**, 162.
14. Scharf, S. J., Horn, G. T., and Erlich, H. A. (1986). *Sciene*, **233**, 1076.
15. Ho, S. N., Hunt, H. D., Horton, R. M., Pullen, J. K., and Pease, L. R. (1989). *Gene*, **77**, 51.
16. Valette, F., Mege, E., Reiss, A., and Adesnik, M. (1989). *Nucleic Acids Res.*, **17**, 723.
17. Horton, R. M., Hunt, H. D., Ho, S. N., Pullen, J. K., and Pease, L. R. (1989). *Gene*, **77**, 61.
18. Saiki, R. K., Gelfand, D. H., Stoffel, S., Scharf, S., Higuchi, R. H., Horn, G. T., Mullis, K. B., and Ehrlich, H. B. (1988). *Science*, **239**, 487.
19. Tindall, K. R. and Kunkel, T. A. (1988). *Biochemistry*, **27**, 6008.
20. Wigler, M., Sweet, R., Sim, G. K., Wold, B., Pellicer, A., Lacy, E., Maniatis, T., Silverstein, S., and Axel, R. (1979). *Cell*, **16**, 777.
21. Wigler, M., Silverstein, S., Lee, L., Pellicer, A., Cheng, Y., and Axel, R. (1977). *Cell*, **11**, 223.
22. Mulligan, R. and Berg, P. (1981). *Proc. Natl Acad. Sci. USA*, **78**, 2072.
23. Nicola, J. F. and Berg, P. C. (1983). In *Teratocarcinoma Stem Cells*, p. 469 (ed. L. M. Silver, G. R. Martin, and S. Strickland). Cold Spring Harbor Laboratory Press, Cold Spring Harbor, NY.
24. Gorman, C. (1985). In *DNA Cloning: A Practical Approach*, Vol. 2 (ed. D. M. Glover), p. 143. IRL Press, Oxford.
25. Neumann, E., Schaefer-Ridder, M., Wang, Y., and Hofschneider, P. H. (1982). *EMBO J.*, **1**, 841.
26. Andreason, G. L. and Evans, G. A. (1988). *BioTechniques*, **6**, 650.
27. Behr, J. P., et al. (1989). *Proc. Natl Acad. Sci. USA*, **86**, 6982.

5

Integrin mutagenesis

A. RESZKA and A. HORWITZ

1. Introduction

Integrins are cell-surface receptors involved in cellular adhesion to the extracellular matrix (ECM), to other cell receptors, and to blood molecules (1–4). In the cell's interior, integrins are connected to the cytoskeleton through a supramolecular linkage. By virtue of these interactions, integrins are implicated in a vast number of functions (*Table 1*).

Table 1. Integrin interactions

Adhesion to matrix	Determination of cell shape
Cell locomotion	Cell–cell adhesion
Platelet aggregation	Binding to blood components
T-cell help	Bacterial invasion
Oncogenesis	Signal transduction
ECM organization	Cytoskeletal organization

Integrin receptors comprise αβ heterodimers. Both α and β subunits have large extracellular, single membrane-spanning, and with the exception of β4 (5, 6), small cytoplasmic domains. The extracellular domains of α subunits are sometimes cleaved into disulfide-linked heavy and light chains, and the β subunit extracellular domains contain cysteine-rich repeats which are thought to be necessary for structure. The β subunits also contain consensus sequences for phosphorylation by pp60^{vsrc} (7), and putative sites for interactions with the cytoskeletal proteins, talin and α-actinin (8, 9).

Specific functions have been attributed to the extracellular and cytoplasmic domains of the integrin receptors (*Table 2*). Substrate specificity is determined primarily by α subunit extracellular domains, which contain multiple calcium-binding motifs. These motifs are responsible for the calcium-dependence of integrin binding interactions. Putative regions for integrin–ECM interactions have been identified by chemical cross-linking techniques. These studies implicate both the α and β subunits. For the $\alpha_{IIb}\beta_3$ and the

Table 2. Potential regions for mutagenesis

Regions	Function
Leader sequence	Surface expression
Extracellular domain	Calcium binding
	ECM binding and specificity
	Heterodimer binding and specificity
	Signal transduction
	Activation
Cytoplasmic domain	Cytoskeletal association
	Phosphorylation
	Signal transduction
	Activation

$\alpha_v\beta_3$ integrins, cross-linked peptides for the β_3 subunits were isolated and sequenced, implicating residues 109–171 in $\alpha_{IIb}\beta_3$ (10) and 61–203 in $\alpha_v\beta_3$ (11). Residues between 134–349 in α_v (33) and 294–314 in α_{IIb} (34) are also implicated. Although the cytoplasmic domains are comparatively small, these are involved in integrin associations with the cytoskeleton and in phosphorylation and regulation.

Mutagenesis offers an approach to identifying amino-acid sequences that contribute to integrin function. The goal of this chapter is to describe approaches and procedures for integrin mutagenesis, explaining tested and useful approaches used in locating active domains and key sequences in these domains. Specifically, the methods for generating mutations, the introduction of mutated DNAs into host cells, and some assays used to test integrin mutants for function will be described. These techniques are also useful in analysing the function of other adhesion molecules.

2. General approach

The first step in any mutagenesis study is the preliminary identification of the regions implicated in the function of interest. Once identified, the amino acids in the region can be mutated, and the mutant integrins can be assayed for changes in function.

There are at least four sources of information that point to a region as involved in a function (*Table 3*). *In vitro* interactions with peptides and chemical cross-linking studies implicate specific regions in direct protein–protein or protein–ligand interactions. Sequence homologies among integrins in the same family (but from different animal species) or different integrin families, or between integrins and other proteins, might identify common (or unique) motifs. Sites targeted for phosphorylation may also be indicative of a

Table 3. Approach to the identification of active domains

- *in vitro* binding studies
- deletion analyses
- homologies
- phosphorylation studies

role in regulation. Finally, key regions can also be identified using deletion mutagenesis coupled with assays for the function of interest.

Once a region has been initially identified, the remaining task is to decide which individual amino acids to mutate. Highly conserved amino acids might be targeted to identify residues involved in a function shared by more than one type of integrin, while non-conserved residues would be targeted when an unique function is assayed.

After targeted regions or residues have been chosen, five major considerations should be addressed before proceeding with the mutagenic study:

- generation of the mutations
- vector for expression of the mutant cDNA
- appropriate host cell system
- introduction of the cDNA into the host cells
- isolation and screening of transfected cells for expression.

2.1 Generation of the mutation

A large number of techniques have been used to generate site-directed mutations. A popular method is oligonucleotide-directed mutagenesis. This technique has been employed successfully to generate termination, internal deletion, and missense mutations in both the α and β integrin subunit cytoplasmic domains (12–16). A recently developed method uses the polymerase chain reaction (PCR). This technique offers rapid generation of mutant DNAs for immediate cloning into expression vectors. PCR can also join two DNAs without restriction endonucleases. The details of PCR mutagenesis are dealt with in another chapter (see Chapter 4).

A variety of protocols for oligonucleotide-directed mutagenesis have been described (17). In our studies, we employ a commercially available kit (Bio-Rad Laboratories, Richmond, California), which is based on the Kunkel method (*Figure 1*) (18). Other mutagenesis kits are also available. They generally contain detailed protocols for oligonucleotide-directed mutagenesis, but do not present information about the design of oligonucleotides. The emphasis of this section focuses on oligonucleotide design.

Integrin mutagenesis

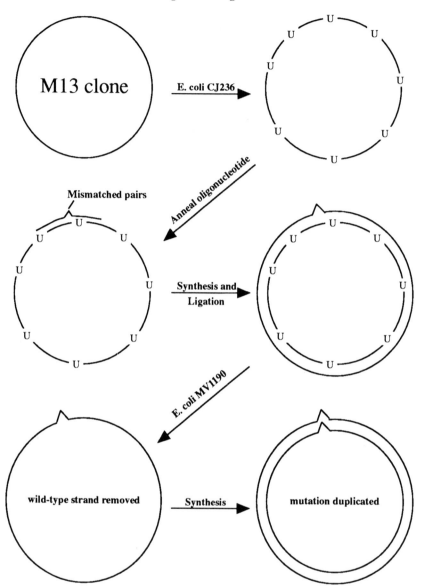

Figure 1. The Kunkel method for oligonucleotide-directed mutagenesis.

Oligonucleotides must include a centrally-located mismatched region, where the mutation is to occur, and a matched region, to either side, where the oligonucleotide can anneal to the wild-type template strand. For generating a single codon change, use the smallest number of mismatches to create the mutation. In the matched region, a good guide is to extend 15–29 nucleotides both 5' and 3' to the mismatchd region (17). We assign matched A and T base pairs a value of two, and C and G base pairs a value of three. We count until a value of at least 30 is attained on either side of the mutation site.

2.1.1 Designing oligonucleotides for deletions

To construct termination deletions, the oligonucleotide should replace the first codon of the deletion with either TAG, TGA, or TAA. It is not necessary to physically remove the sequence from the cDNA, since these stop codons prevent translation beyond their location in the mRNA. For internal deletions, one must physically remove the coding region for the sequences deleted. A simple method involves the generation of novel restriction sites in the cDNA at the borders of the deletion. The deletion itself is actually generated by cutting at the two restriction sites and religating, thus eliminating the intervening sequence. It is best to try to match the restriction endonuclease site to the codons forming the borders of the deletion. One should try to incorporate unique restriction sites to avoid the use of technically difficult partial digestions in forming the actual deletion. This eliminates the possibility of an altered amino acid residue causing errant results. Additionally, one must take care to preserve the reading frame downstream of the mutation.

Problems with protein folding can be avoided by designing deletions to remove gene exons, which typically represent individual domains. Until recently, little was known about the intron–exon structure of integrin genes. However, the extracellular, transmembrane, and cytoplasmic domains are easily identified from cDNA sequence data.

2.1.2 Designing oligonucleotides for point mutations

In designing a point mutation, the choice of amino acid replacements is most important. Substitutions for targeted amino acid residues generally fall into three categories:

(a) One choice is to replace all targeted residues with alanine. Alanine has a small R-group, but unlike glycine, alanine is less likely to alter secondary structure.

(b) Some choose to replace targeted residues with residues that maintain size, shape, charge, polarity, or hydrophobicity. *Table 4* lists some typically targeted amino acids and possible substitutions, including acidic substitutions designed to mimic phosphorylation and alanine substitutions for residues with hydrophobic rings. The amino acids listed are offered as a guide, but do not represent all possible substitutions. Except where noted, these substitutions have been chosen to maintain the general size and shape of the targeted residue, changing only hydrophobicity and charge. Because it is often difficult to predict the effect of one mutation, one might use multiple substitutions at the same site.

Integrin mutagenesis

Table 4. Possible amino acid substitutions for point mutations

Amino acid target	Substitutions
Asp, Glu	Asn, Gln, Ile, Leu, Val
Asn, Gln	Ile, Leu, Val
Lys, Arg, His	Gln, Ile, Leu, Met
Ser, Thr	Cys, Leu, Met, Val, Asp*, Glu*
Pro	Ala, Gly
Tyr	Ala†, Phe, Asp*, Glu*
Phe, Trp	Ala†, Trp, Phe

* Denotes substitutions used to mimic phosphorylation states.
† Denotes substitution to remove hydrophobic ring.

(c) Finally, others choose to replace targeted residues randomly. Multiple mutations at the same site can be made with more than one oligonucleotide, each designed to create one mutation. Alternatively, a degenerate oligonucleotide has the potential for up to 20 substitutions at the same site. The advantage of degenerate oligonucleotides is that they are cheaper; however, they also generate a background of undesirable (too conservative or non-conservative) mutations.

2.2 Expression vectors

Careful choice of an expression vector will ensure isolation of transfected cells with high protein expression levels. A wide variety of vectors have been developed for expression in eukaryotic cells and many are commercially available. A good vector should include:

(a) A strong promoter for high integrin expression levels. Some commercially available vectors include the RSV, SV40, mouse mammary tumour virus (MMTV), thymidine kinase, or metallothionein I promoters.

(b) A marker for selection of transfected cells. Available selectable markers include: the neomycin or hygromycin resistance genes, or the *Escherichia coli* guanine phosphoribosyltransferase (*gpt*) gene.

(c) An *E. coli* origin of replication and ampicillin resistance gene for isolation of large quantities of plasmid from *E. coli* cultures.

(d) The ability to generate virion particles. If it is difficult to introduce an expression vector into a specific host cell, or if a high percentage of transfectants is desired, virion particles can greatly enhance the efficiency of delivery. *Table 5* lists factors to consider when choosing or designing an expression vector.

Table 5. Factors to consider in choosing an expression vector

- compatibility between promoter and type of host cell
- presence of restriction sites for easy insertion of cDNA
- presence of marker for selection of transfected cells
- presence of sequences for propagation and selection in *E. coli*
- ability to generate virion particles for enhanced delivery to host

2.3 Host cell choice

The first decision in choosing a host cell involves deciding between primary and immortal cells. Other considerations include how the introduced integrin will function in the host cell, whether or not other receptors exist in the cell that have similar or identical functions, and how easy it is to introduce the foreign DNA into the cell (see *Table 6*).

Primary cells are often chosen since they best approach an *in vivo* system. While primary cells may represent the most accurate system for examining a particular problem, they are typically difficult to transfect, and they usually have finite life spans in culture. To circumvent the problem, transient expression assays are often employed. For this, it is important to enhance the delivery of the vector DNA into the cells so that more material is available for assay. This is often done with retrovirion packaging systems, which incorporate the plasmid into virus particles as RNA copies. By infecting cells with a high multiplicity of infection, a larger percentage of the population will express the introduced integrin.

Using a cell line offers the advantage of long-term propagation. This is important for selection with antibiotics. For example, G418 usually takes 7 to 9 days to kill NIH 3T3 cells. Cell lines are also typically easier to transfect than primary cells. Furthermore, some cell lines were designed to support episomal replication of vectors bearing the SV40 origin of replication (19). For example, COS-7 cells express the SV40 T antigen, which initiates episomal replication at the SV40 origin of replication. Vectors that do not contain this origin of replication, or vectors expressed in cells lacking the SV40 T antigen, do not replicate, and, in fact, must integrate into the host chromosome for stable replication of the vector DNA.

Table 6. Factors to consider in choosing a host cell

- ease of introduction of DNA
- presence/absence of host receptors with similar substrate specificity
- immunological distinction between host and introduced integrin
- ability of host to utilize introduced integrin for desired function

2.4 Introduction of the cDNA into the host cells

The American Type Culture Collection is a good source of cells and their catalogue is a useful source of information regarding host cell attributes (20).

Transfection is a method of introducing a cDNA into a host cell. A variety of transfection techniques are available, including: calcium phosphate, lipofection, and electroporation (17). We have found that calcium phosphate precipitation works with most cell lines (21). Calcium phosphate precipitation protocols have many variations that can be used to enhance their efficiency. These include the use of glycerol shock (22) and chloroquine treatment (23). We commonly use the glycerol shock technique to transfect NIH 3T3 cells. *Protocol 1* contains a method for introduction of vector DNA into host cells by calcium phosphate precipitation followed by glycerol shock.

Protocol 1. Calcium phosphate precipitation and glycerol shock

Materials

- 15 μg host carrier DNA (or other carrier DNA)[a]
- 0.25–10 μg vector DNA (to be determined for each vector and cell type)
- 2 × Hepes-buffered saline (HeBS) (0.28 M NaCl, 10 mM KCl, 14 mM Na_2HPO_4, 12 mM dextrose, and 40 mM *N*-2-hydroxyethylpiperidine-*N'*-2-ethane sulfonic acid [Hepes]–HCl), pH 7.0–7.1, sterile[b]
- 1 × HeBS, 15% glycerol, pH 7.0–7.1 sterile
- 0.5 × TE pH 7.5 (5 mM Tris, 0.5 mM EDTA)
- 2.5 M $CaCl_2$, sterile
- sterile distilled water
- DMEM or suitable media
- polystyrene tubes

Methods

1. Plate cells at $3–6 \times 10^5$ cells/ 6-cm dish 20–24 h prior to addition of precipitate.
2. Suspend vector and carrier DNA in 75 μl 0.5 × TE.
3. Add 150 μl sterile distilled water and 25 μl sterile 2.5 M $CaCl_2$.
4. Transfer 250 μl 2 × HeBS to polystyrene tube.
5. With gentle agitation, add the DNA solution slowly to the 2 × HeBS, a fine precipitate is desired.
6. Incubate mixture at room temperature (RT) 30–45 min.
7. Pipette to mix and add the precipitate to the cells, swirl or pipette up and down to distribute DNA over cells.[c]

8. Incubate 37°C 4 h.
9. Aspirate media.
10. Rinse cells once with serum-free media and aspirate.
11. Add 1 ml 1 × HeBS, 15% glycerol to cells.
12. Incubate for 2 min at RT.[d]
13. Add serum-free media with swirling to cells, then aspirate.
14. Rinse once with serum-free media, then aspirate.
15. Add complete media and return to incubator.[3]

[a] Carrier DNA increases transfection efficiency, but can be omitted.
[b] The pH for both HeBS buffers should be adjusted on the day of their use.
[c] Calcium phosphate precipitate can be added to cells after removing the media. Aspirate media, add precipitate, incubate for 20 min at RT, then add media back to the dish.
[d] Incubation times for glycerol shock can vary from 1–3 min.
[e] Cells should be incubated for 24–48 h after transfection, before selection with antibiotics.

2.5 Isolation and screening for expression

Two major approaches can be used to generate cells expressing integrin on their surface. Clone isolation generates cells derived from one progenitor. Flow cytometry and panning generate heterogeneous cell populations using surface expression as criterion for selection. Both techniques use antibiotics to enhance the number of integrin-expressing cells.

2.5.1 Use of selectable markers to eliminate untransfected cells

The most popular methods of generating populations of tranfectants use antibiotics such as G418 and hygromycin to eliminate untransfected cells. G418 resistance in is conferred by the bacterial transposon (Tn5) neomycin resistance (aminoglycoside phosphotransferase) gene (24), while hygromycin resistance is encoded by the *E. coli* hygromycin B phosphotransferase gene (25).

Selecting transfected cells with G418 or hygromycin requires incubation for at least 24 to 48 h, prior to selection in order to allow resistance gene expression. Selection time and antibiotic concentrations will depend on cell type. If one is using a cell with unknown antibiotic resistance, test for cell death using a number of concentrations of antibiotic over a time-course of 1–2 weeks. In the actual transfection experiment, untransfected cells will detach from the dish in 7 to 14 days (media should be changed every 2–3 days).

2.5.2 Clone isolation

Two approaches to clone isolation are generally used: limiting dilution and colony isolation. Limiting dilution involves diluting and seeding cells into multiwell plates such that each well receives one or less cell. In colony isolation, the clone grows as hundreds of cells in a colony which is transferred to the well. Since neither technique guarantees that the cells isolated express the integrin cDNA, all isolates must be screened prior to further use.

The limiting dilution procedures works well for cells that cannot be grown in contact with each other. Cells are grown at subconfluency in antibiotic media to enrich the number of transfected cells. Cells are then collected by trypsinization and seeded into multiwell plates at less than 1 cell/well. After propagation, cells are assayed for integrin expression.

Since limiting dilution is time-consuming, many choose to allow transfected clones to form under antibiotic selection as colonies; however, the cells must be able to grow in contact with each other. Initial cell densities must be adjusted when beginning antibiotic selection, such that colonies form with a 3–10 mm separation from each other. This ensures that a colony is isolated without contamination from another. After the colonies form on the Petri dish, they are isolated using cloning rings, then are transferred to individual wells in a multiwell plate. Like clones developed by limiting dilution, these are then screened for integrin expression. However, this technique leads to rapid screening for integrin expression, since the original cell number seeded into each well is hundreds rather than one.

Protocol 2. Use of cloning rings to isolate colonies

Materials
- Versene solution:

Chemical	g/litre
NaCl	8.2 g
KCl	0.2 g
Na_2HPO_4	2.17 g
KH_2PO_4	0.2 g
EDTA	0.2 g

  Bring to 1 litre with water and adjust to pH 7.4 with HCl.
- forceps
- glass Petri dish
- high-vacuum grease (Corning)
- yellow pipette tips (any brand)
- 0.25% trypsin in versene (TRED)
- growth medium (varies from cell to cell)[a]

Methods
1. Spread a thin layer of vacuum grease over the bottom of the glass Petri dish. Sterilize by autoclaving.
2. Cut the bottoms off of yellow pipette tips 5–8 mm from the end. Sterilize by autoclaving.[b]

3. Aspirate media from culture dish, (cell colonies should be well separated from each other) and circle designated colonies on the outside of the dish with a marking pen.
4. With sterile forceps, dip the smooth end of a sterile cloning ring into the sterile vacuum grease and press over the circled colony.
5. Add 60 μl TRED to each ring, incubate for 1–2 min at room temperature.
6. Pipette up and down to loosen cells, and transfer to separate Petri or multi-well cluster plate containing growth media and the selective antibiotic.[c]

[a] For NIH 3T3 cells: Dulbecco's modified Eagle's medium (DMEM), supplemented with 10% calf serum and selective antibiotic (G418 can be used between 200–1000 μg/ml).
[b] For isolating clones, we have found that yellow pipette tips (200 μl size) work well as cloning rings. Since these are disposable, there is no need to clean the grease off them after they are used. Cut the bottoms off, using a razor or a hot wire. The factory-smooth end is flat enough to form a tight seal to the Petri dish bottom after being dipped into the sterile vacuum grease.
[c] To ensure that cloned cells maintain the expression vector, grow cells in the presence of the antibiotic used for selection. For a vector encoding G418 resistance in NIH 3T3 cells, we use G418 at 1 mg/ml for selection and 250 μg/ml after cloning.

2.5.3 Flow cytometry

Both panning and flow cytometry isolate cells expressing integrin on their surface. In panning, an anti-integrin antibody is conjugated to a plastic well and cells expressing integrins on their surface attach to the antibody and untransfected cells remain in solution (see Chapter 3). In flow cytometry, FITC-conjugated antibodies are attached to the cells, and the cells are sorted, based on relative surface expression levels. We prefer flow cytometry because it also quantitates surface expression levels.

For flow cytometry, one skips the cloning step, and grows the transfectants as a mixed culture. Cells are immunostained live with anti-integrin antibodies conjugated to FITC (*Protocol 3*) (26). Fluorescein isothiocynate (FITC) can be directly conjugated to the primary antibody, or to a second antibody that recognizes the primary. The cells are analysed and collected using a flow cytometer. Specific protocols for analysis of labelled cells vary depending on the hardware used. Typical data obtained from an EPICS™ flow cytometer using MDADSII™ software (Coulter Electronics, Inc., Miami Lakes, Florida) are presented in *Figure 2*, which shows the surface expression profiles for three transfected cell populations. For this assay, cells, with fivefold higher surface expression (**a, b, c**) relative to mean background fluorescence (**d**), were sorted from non-clonal populations. After sorting, cells were returned to culture dishes, expanded, and sorted again. After one round of sorting, an increase in mean surface expression is observed (**a′, b′,** and **c′**). Successive sorting generates populations that maintain high average expression levels.

Integrin mutagenesis

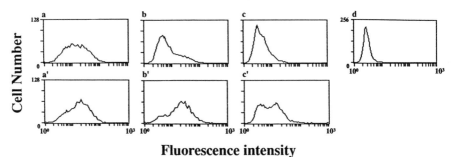

Figure 2. EPICS™ flow cytometry profiles for avian integrin β_1 subunit-transfected NIH 3T3 cells. (**a** and **a'**) Wild-type, (**b** and **b'**; **c** and **c'**) two cytoplasmic domain mutants, and (**d**) untransfected NIH 3T3 cells were labelled with the avian β_1-specific monoclonal antibody, W1B10. Flow cytometer profiles show surface expression levels for 5000 cells before (**a, b,** and **c**) and after (**a', b',** and **c'**) cells were sorted using a window fivefold over mean background fluorescence (**d**).

Protocol 3. Cell preparation for flow cytometry

Materials (0.02% NaN$_3$ added to each solution)

- CMF–HH (calcium, magnesium-free PBS–Hepes, Hanks'):

Chemical	g/litre
KCl	0.4
KH$_2$PO$_4$	0.06
NaCl	8.0
NaHCO$_3$	0.35
Na$_2$HPO$_4$*7H$_2$O	0.09
glucose	1.0
Hepes	4.7

  Bring to 1 litre with water and adjust to pH 7.4 w/NaOH (sterilize)

- BB (blocking buffer): CMF–HH, 2% BSA (sterile)
- CMF–HH–EDTA:EDTA at 0.2 g/litre (sterile)

Methods

1. Grow cells in one or two 10-cm dishes to 70–85% confluency.
2. Wash twice with CMF–HH.
3. Add 0.5–1.0 ml CMF–HH–EDTA.
4. Incubate at 37°C for 20–30 min.
5. Collect by pipetting up and down (to create single-cell suspension) and centrifuge in microfuge (Eppendorf 5415) at 1000 xg, 2 min, RT.
6. Resuspend in 1 ml BB containing antibody at 20–30 μg/ml.

7. Incubate at 4°C for 30 min; invert every 10 min.
8. Centrifuge and aspirate.
9. Resuspend in 1 ml BB, repeat step 8.
10. Resuspend in 1 ml BB containing FITC-conjugated secondary antibody at 1:200 (dilution required may vary).
11. Incubate at 4°C for 30 min, invert every 10 min.
12. Centrifuge and aspirate.
13. Resuspend in 1 ml BB, repeat steps 12 and 13 (twice).
14. Resuspend in CMF–HH so that cells are at approx. 10^6/ml.

3. Applications of integrin mutagenesis

Mutations in the integrin α and β cytoplasmic domains have been very useful for identifying residues that mediate focal adhesion localization, cellular adhesion, and integrin activation. Studies of the integrin extracellular domains have not yet been reported. This section focuses on the integrin cytoplasmic domains. Three different mutagenesis studies will be discussed, some of the functional assays described, and mutagenesis data presented. The interpretation of the results of these experiments is presented to indicate how these experimental methods might be used for the analysis of other molecules.

3.1 Focal adhesion localization of mutant integrins

The most extensive studies have been targeted towards identifying avian integrin $β_1$ subunit cytoplasmic sequences that contribute to localization in focal adhesions of fibroblasts. The common assay in all of these studies is immunostaining of transfected NIH 3T3 cells to detect integrins in focal adhesions.

For staining, we recommend that cells are first sorted in a flow cytometer (Section 2.5.2), using a narrow window of surface expression. This assures that the different mutant-expressing cells have equivalent amounts of integrins on their surface. The cells are allowed to recover overnight, plated on fibronectin substrates, and allowed to spread for 4–6 h. The cells are next labelled by indirect immunofluorescence as described in *Protocol 4*. Finally, immunostained cells are examined using a fluorescence microscope.

Protocol 4. Live-cell staining for integrin

Materials
- PBS
- primary antibody directed against integrin subunit
- FITC or rhodamine-conjugated antibody directed against primary antibody

Protocol 4. *Continued*
- formaldehyde
- goat serum (or equivalent serum matched to animal from which secondary antibody obtained)
- glycine

Methods
1. Rinse cells three times with PBS.
2. Add primary anti-integrin antibody at 10–50 μg/ml in PBS, containing 5% goat serum; incubate 20 min at RT.
3. Rinse 10–12 times with PBS.
4. Fix with 3% formaldehyde in PBS 15 min at RT.
5. Add 0.15 M glycine in PBS 7.5 min at RT (repeat once).
6. Rinse three times with PBS.
7. Block with 5% goat serum 15–60 min at RT.
8. Add secondary (Rhodamine or FITC-conjugated) antibody (1:150–1:200 dilution in PBS, containing 5% goat serum) for 20 min at RT.
9. Rinse 10–12 times with PBS.
10. Mount and observe, using fluorescent microscope.

Focal adhesion staining intensities of the mutants are determined, relative to wild-type, by comparing fluorescence micrographs showing focal adhesion staining (*Figure 3*). Four localization phenotypes are used: normal, reduced, trace, and null. The mouse NIH 3T3 cells express the murine β_1 subunit, in addition to the transfected chicken subunit. Therefore, cells expressing avian β_1 subunits with reduced or null localization phenotypes, show normal cell morphology. However, antibodies directed against the avian β_1 subunit do not cross-react with the murine β_1 subunit, therefore untransfected cells do not show any focal adhesion staining.

3.1.1 Localization of β_1 subunit cytoplasmic deletion mutants

The initial cytoplasmic mutation studies used a series of termination codon-derived deletion mutations generated in the avian β_1 subunit (12–14). Small deletions of four or five C-terminal amino acids resulted in trace localization to focal adhesions, while larger deletions extending into or beyond the central portion of the (41 amino acid) cytoplasmic domain resulted in a null localization phenotype. Two internal deletions were also created: one ten-residue deletion adjacent to the transmembrane domain and another twenty-residue deletion in the central region of the cytoplasmic domain. Both deletion mutants resulted in trace localization to focal adhesions.

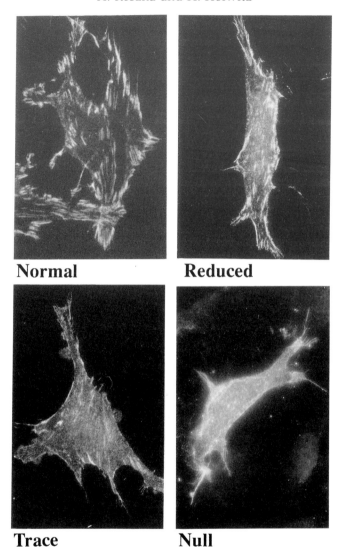

Figure 3. Focal adhesion staining of wild-type β_1 integrin and four cytoplasmic domain mutants.

Analysis of these data was as follows:

(a) The β_1 subunit cytoplasmic domain contains key sequences for localization to focal adhesions.

- Larger β_1 subunit termination deletions resulted in null localization to focal adhesions.

Integrin mutagenesis

- Resident α subunit sequences were not altered in these studies, yet alone, were unable to mediate localization to focal adhesions.

(b) The β₁ subunit cytoplasmic domain contains multiple signals or sequences for localization to focal adhesions

- Individual deletions adjacent to the membrane, in the central region, and at the C-terminus, resulted in trace localization phenotypes, suggesting that localization sequences exist outside the deleted areas
- Null localization phenotypes were seen only when the central and C-terminal regions were removed by larger termination deletions.

3.1.2 Localization of integrin point mutants

The next step used to study the β₁ subunit cytoplasmic domain involved point mutations. Since the β₁ subunit cytoplasmic domain deletions did not narrow the field of potential areas containing localization signals, it was necessary to examine all regions. Using the guide-lines set in Section 2.1.2, residues were replaced with either isosteric substitutions, or those that maintained charge or hydrophobicity. A summary of β₁ subunit mutations that reduced or did not alter focal adhesion staining intensities are shown in *Figure 4*. Mutations that did not reduce focal adhesion localization are drawn above the wild-type sequence, while those that did are drawn below. Focal adhesion reducing mutations cluster into three regions, termed cyto-1: D̲RREF̲AKF̲EKE̲, cyto-2: NPI̲YKS, and cyto-3: NPKY̲.

Analysis of β₁ subunit cyto-1 mutation data was as follows:

Cyto-1: DRREFAKFEKE: No previous data regarding cyto-1 were available before mutagenesis, therefore, descriptions of this region are all based on mutagenic data.

(a) Only four of eleven residues are required for function:

- Replacement of the aspartate, both phenylalanines, and the last glutamate (D̲RREF̲AKF̲EKE̲) resulted in reduction of focal adhesion localization.
- Equivalent substitutions of adjacent residues did not alter focal adhesion localization phenotypes.

Figure 4. Locations for and summary of avian β₁ integrin point mutations and focal adhesion localization phenotypes. *Note*: The Tyr to Glu substitution at this site reduced localization more than the Tyr to Ala substitution did.

(b) Cyto-1 is integrin β subunit-specific and highly conserved:
- A database search for proteins bearing cyto-1 found only integrin β subunit matches (27).
- The four key amino acids are highly conserved among the different β subunit families (in the $β_6$ subunit, the first phenylalanine is replaced with alanine).

(c) The four key amino acids of cyto-1 form a signal on one face of an α helix:
- The eleven residues form an α helix by computer-generated predictions (DNAstar, Madison, Wisconsin) using two different algorithms (28, 29).
- All amino acid substitutions are predicted to preserve α helix and β sheet, structures (28).
- The positions of four key amino acids show a periodicity of 3–4 residues, and therefore cluster to one side when drawn on a helical wheel (30).

Conclusion: Cyto-1 is an unique β subunit focal adhesion localization signal that forms on one face of an α helix.

The same approach can be used to examine both cyto-2 and cyto-3. Using this type of analysis, the following conclusions are made: Cyto-2 forms an obligatory tight turn and may form a signal that contributes to localization to focal adhesions. Integrins may be negatively regulated at or adjacent to cyto-2 by tyrosine or serine phosphorylation. Cyto-3 contains an NXXY motif which contributes to localization in focal adhesions.

3.2 Assay of integrin mutants for cellular adhesion

Three studies have described the adhesion of cells expressing integrins with mutated cytoplasmic domains (13, 15, 31). Two studies involved avian $β_1$ subunit adhesion to fibronectin and laminin. The third examined the interaction between the $α_Lβ_2$ integrin, which mediates cell–cell interactions between leucocytes and cells bearing the ICAM receptor, and purified ICAM. The assays for cellular adhesion vary for each adhesion receptor and cell type, and specific methods are covered in Chapter 1. However, there are a number of considerations that are applicable to most adhesion systems (*Table 7*).

One must first develop a means to assay only the mutant integrin for cellular adhesion by eliminating the contribution of host integrins or other receptors with similar functions. A first step to control for this is to assay an untransfected cell population for adhesion. If untransfected cells adhere to the substrate, then further steps must be taken to isolate the mutant integrin in the adhesion system. If the cell has receptors that are interfering with the assay, they often can be functionally blocked using antibodies. Cells and antibodies are incubated prior to plating on the adhesive substrate. If the cells

Table 7. Factors involved in integrin adhesion assays

- presence of other adhesion molecules
- concentrations and purity of substrate
- time-course for cell attachment and spreading
- need to block synthesis and secretion of:
  host receptors that may bind substrate
  other unrelated ECM or adhesive substrate
- means to remove cells that adhere non-specifically

secrete more receptors during the process of adhesion and spreading, one must either block protein synthesis and secretion, or limit the time allowed for adhesion to occur.

Another consideration involves impure preparations of ECM or the secretion of ECM by the cell. One simple method to test for both of these problems is to include a control assay where the integrin of interest is functionally blocked with an antibody. If one runs the adhesion assay with this integrin blocked, none of the cells should adhere to the substrate. If the number of adherent cells rises above normal background, then the purity of and/or the synthesis of other adhesive substrates should be examined.

3.2.1 Cellular adhesion of avian β_1 subunit mutants

Two types of avian β_1 integrin mutants have been assayed for adhesion to ECM substrates. One avian β_1 deletion mutant has been assayed for cellular adhesion to fibronectin (31). The integrin mutant, 765t, had the C-terminal 39 amino acid residues removed by the incorporation of a terminator codon in place of the 765th codon. In transfected NIH 3T3 cells, this mutant integrin was null for localization to focal adhesions. In order to assay for adhesion, the murine β_1 subunit was blocked with a polyclonal antibody (32) prior to addition of the cells to fibronectin or laminin-coated microwells. Although the wild-type avian β_1 subunit did function in cellular adhesion, the 765t mutant did not. This assay provided the first evidence that β_1 subunits are functionally interchangeable between animal species and that the β_1 subunit cytoplasmic domain is required for cellular adhesion via β_1 integrins.

A chimeric mutant, $\beta_{1/3}$, has also been assayed for adhesion to laminin (31). This chimera was made by connecting the β_1 extracellular domain to the β_3 transmembrane and cytoplasmic domains. The β_1 subunit can associate with the α_1, α_2, α_3, α_4, α_5, α_6, and α_v subunits, while the β_3 subunit only associates with the α_{IIb} and α_v subunits (2). In the NIH 3T3 cells, $\beta_{1/3}$ associated with β_1-specific α subunits, demonstrating that heterodimer formation was determined by the extracellular domain. To assay for adhesion, a similar approach was used to block the resident β_1 subunit with a polyclonal antibody prior to adhesion to laminin. This mutant adhered with a normal phenotype, suggesting that the β_1 and β_3 subunits form similar cytoskeletal linkages.

3.2.2 Cellular adhesion of human $\alpha_L\beta_2$ deletion mutants to ICAM

Recently, termination deletion mutants of the $\alpha_L\beta_2$ integrin cytoplasmic domains were assayed for cellular adhesion to purified ICAM (15). Although the $\alpha_L\beta_2$ integrin is normally found on leukocytes, both subunits were co-expressed in COS-7 cells, in which they were able to mediate cellular adhesion to purified ICAM. Because untransfected COS-7 cells do not normally express $\alpha_L\beta_2$ or adhere to ICAM, no methods were needed block resident receptors. Using a series of termination codon-derived deletion mutations of both the α_L and β_2 subunits, the importance of the cytoplasmic domains of both subunits in cell adhesion was determined. Deletions of the β_2, but not the α_L, cytoplasmic domain resulted in eliminated or impaired binding to ICAM. In this mutagenic study, both the ubiquitous nature of the integrin cytoskeletal linkages and the importance of the β_2 subunit for cellular adhesion were demonstrated.

3.3 Activation of human $\alpha_{IIb}\beta_3$ deletion mutants

Deletion studies on $\alpha_{IIb}\beta_3$ have determined a function for the α_{IIb} subunit in integrin activation (16). The $\alpha_{IIb}\beta_3$ integrins are present on platelets where they mediate aggregation when activated. In this study, α_{IIb} and β_3 cDNAs were co-expressed in CHO cells, which express neither of these receptors normally. Wild-type, deletion mutants and an α_{IIb}/α_5 chimera, exchanging the α_{IIb} cytoplasmic domain for one from α_5, were assayed for the activation state of the integrin. Flow cytometry (Section 2.5.2) was used as an analytical assay for integrin activation, using an antibody, PAC1, which recognizes a β_3 integrin epitope that is exposed only after activation. A normal (inactive) state was observed for wild-type and β_3 deletion integrins, but the α_{IIb} deletion and α_{IIb}/α_5 chimeric mutants were constitutively activated. This flow cytometry data was correlated directly with assays for cellular aggregation, showing that the α_{IIb} deletion and α_{IIb}/α_5 chimeric mutants were also constitutively activated for cellular aggregation. This mutagenic study provided the first evidence that regulation of $\alpha_{IIb}\beta_3$ integrin activation is mediated through the α_{IIb} subunit cytoplasmic domain.

References

1. Hemler, M. E. (1990). *Annu. Rev. Immunol.*, **8**, 365.
2. Albeda, S. M. and Buck, C. A. (1990). *FASEB*, **4**, 2868.
3. Springer, T. A. (1990). *Nature*, **346**, 425.
4. Dustin, M. L. and Springer, T. A. (1991). *Annu. Rev. Immunol.*, **9**, 27.

5. Suzuki, S. and Naitoh, Y. (1990). *EMBO J.*, **9**, 757.
6. Hogervorst, F., Kuikman, I., von dem Borne, A. E. G. Kr., and Sonnenberg, A. (1990). *EMBO J.*, **9**, 765.
7. Hirst, R., Horwitz, A., Buck, C., and Rohrschneider, L. R. (1986). *Proc. Natl Acad. Sci. USA*, **83**, 6470.
8. Tapley, P., Horwitz, A., Buck, C., Burridge, K., Duggan, K., and Rohrscheider, L. (1989). *Oncogene*, **4**, 325.
9. Otey, C. A., Pavalko, F. M., and Burridge, K. (1990). *J. Cell Biol.*, **111**, 721.
10. D'Souza, S. E., Ginsberg, M. H., Burke, T. A., Lam, S. C.-T., and Plow, E. F. (1988). *Science*, **242**, 91.
11. Smith, J. W. and Cheresh, D. A. (1988). *J. Biol. Chem.*, **263**, 18726.
12. Solowska, J., Guan, J.-L., Marcantonio, E. E., Trevithick, J. E., Buck, C. A., and Hynes, R. O. (1989). *J. Cell Biol.*, **109**, 853.
13. Hayashi, Y., Haimovich, B., Reszka, A., Boettiger, D., and Horwitz, A. (1990). *J. Cell Biol.*, **110**, 175.
14. Marcantonio, E. E., Guan, J.-L., Trevithick, J. E., and Hynes, R. O. (1990). *Cell Reg.*, **1**, 597.
15. Hibbs, M. L., Xu, H., Stacker, A., and Springer, T. A. (1991). *Science*, **251**, 1611.
16. O'Toole, T. E., Mandelman, D., Forsyth, J., Shattil, S. J., Plow, E. F., and Ginsberg, M. H. (1991). *Science*, **254**, 845.
17. Sambrook, J., Fritsch, E. F., and Maniatis, T. (ed.) (1989). *Molecular Cloning: A Laboratory Manual.* Cold Spring Harbor Laboratory Press, Cold Spring Harbor, NY.
18. Kunkel, T. A., Roberts, J. D., and Zakour, R. A. (1987). *Methods in Enzymology*, Vol. 154 (ed. R. Wu), p. 367. Academic Press, Orlando, Florida.
19. Gluzman, Y. (1981). *Cell*, **23**, 175.
20. *American Type Culture Collection Catalogue of Cell Lines and Hybridomas* (6th edn) (1988).
21. Graham, F. L. and van der Eb, A. J. (1973). *Virology*, **52**, 456.
22. Frost, E. and Williams, J. (1978). *Virology*, **91**, 39.
23. Luthman, H. and Magnusson, G. (1983). *Nucleic Acids Res*, **11**, 1295.
24. Southern, P. J. and Berg, P. (1982). *J. Mol. Appl. Gen.*, **1**, 327.
25. Gritz, L. and Davies, J. (1983). *Gene*, **25**, 179.
26. George-Weinstein, M., Decker, C., and Horwitz, A. (1988). *Devel. Biol.*, **125**, 34.
27. Pearson, W. R. and Lipman, D. J. (1988). *Proc. Natl Acad. Sci. USA*, **85**, 2444.
28. Chou, P. Y. and Fasman, G. D. (1978). *Adv. Enzymol.*, **47**, 45.
29. Guarnier, J., Osguthorpe, D. J., and Robson, B. (1978). *J. Mol. Biol.*, **120**, 97.
30. Schiffer, M. and Edmundson, A. B. (1967). *Biophys. J.*, **7**, 121.
31. Solowska, J., Edelman, J. M., Albeda, S. M., and Buck, C. A. (1991). *J. Cell Biol.*, **114**, 1079.
32. Akiyama, S. K. and Yamada, K. M. (1987). *J. Biol. Chem.*, **262**, 17536.
33. Smith, J. W. and Cheresch, D. A. (1990). *J. Biol. Chem.*, **265**, 2168.
34. D'Souza, S. E., Ginsberg, M. H., Matseuda, G. R., and Plow, E. F. (1990). *J. Biol. Chem.*, **265**, 3440.

6

Isolation of intercellular junctions

ELLIOT L. HERTZBERG, BRUCE R. STEVENSON,
KATHLEEN J. GREEN, and SHOICHIRO TSUKITA

1. Introduction—Elliot L. Hertzberg

Early electron microscopic studies revealed the existence of a number of distinct types of intercellular junctions. Specific physiological properties of a cell or tissue could often be correlated with the presence of a particular type of junction, suggesting very different functions for each type. Some types of cell contacts, the desmosomes and adherens junctions, appeared to tether different elements of the cell's cytoskeleton to the plasma membrane while forming adhesion points between cells. Others were proposed to provide a barrier to paracellular diffusion of solutes (tight junctions) or facilitate intercellular communication (gap junctions). Roles were hypothesized for these junctions in tissue modelling, and in regulation of cell function and cell proliferation.

The ability to isolate intercellular junctions has led to the development of more detailed molecular models of these junctions—and to many surprises. For gap junctions, a family of structural proteins, the connexins, has now been identified, presumably with individual properties which influence the activity of intercellular channels. Tight junctions are now known to not only form a diffusion barrier for solutes, but for plasma membrane proteins and lipids, thereby creating discrete domains within the cell's limiting membrane. Adhesive junctions are now being found to share some proteins and overlap in their interactions with cell adhesion molecules, a feature in common with contact points between cells and substrata and the extracellular matrix.

While complexity has increased, common principles are being found. And the ability to test the hypothetical roles of these intercellular junctions in cellular and developmental processes is now, with the isolation and characterization of intercellular junctions, yielding to experimental analysis.

2. Rodent liver gap junctions

2.1 Distribution and function of gap junctions

Gap junctions, also referred to (with increasing rarity) as the nexus or maculae communicans, were originally observed in thin section images of heart (1), and smooth (2) muscle, as well as in Mauthner cells in goldfish brain (3). In addition to the close appositional surfaces of adjacent cells' plasma membranes, the glancing section in the Mauthner cell micrographs revealed a polygonal lattice of subunits containing an electron dense spot, penetrated by heavy metal stain, in the centre. This spot is now known to be the site of the intercellular channel which provides intercellular communication by mediating direct exchange of low-molecular-weight (about 1000 daltons in vertebrates) hydrophilic molecules between contacting cells. Gap junctions may be found between all cell types except for mature skeletal muscle and mature circulatory cells. In smooth and heart muscle, as well as in neurones which can be electronically coupled via gap junctions, impulse transmission occurs as a result of ion flux through gap junction channels. In these and other tissues and cell types, gap junctional communication is thought to play a role in the integration of tissue function and transmission of signals which might regulate differentiation and cell growth.

The rodent connexin family of gap junction proteins consists of at least eleven (D. Paul, personal communication) highly related proteins, most of which were initially identified by DNA cloning. Several common structural motifs have been found among these connexins. Each contain four transmembrane segments with the amino and carboxy termini located at the cytoplasmic surface of the junction. The two extracellular loops and the transmembrane segments are relatively conserved, while the cytoplasmically disposed sequences show the greatest variability. This is consistent with observed ability of cells containing different connexins to establish communication while maintaining distinct regulatory patterns for each cell's junction proteins. Connexins, based on Western and Northern blot analyses, have a greater or lesser degree of abundance in different cell types. Thus far, most attention has focused on connexin32 and connexin26, which were originally identified as constituents of isolated liver gap junctions (4–6), and connexin43, originally found as a component of isolated myocardial gap junctions (7–9).

The successful isolation of gap junctions from rodent liver was an outgrowth of the observation of Benedetti and Emmelot (10) that electron micrographs of the residue from detergent treatment of isolated liver plasma membranes indicated an enrichment for gap junctions. Several investigators developed purification procedures involving treatment of isolated liver plasma membranes with detergents such as deoxycholate, Triton X-100 or *N*-lauroyl sarcosinate (Sarkosyl) with subsequent purification based upon physical separation (i.e. differential centrifugation and sucrose density gradient centrifugation). The most confusing aspect of the studies from this period was that each investigator appeared to be isolating different protein(s). This arose, in part, because different laboratories were using different proteases as 'aids' for tissue dissociation and that there was relative homogeneity in the contaminating proteins in the material obtained by the different procedures. The initial isolation of the proteins now called connexins in reasonably homogeneous form as constituents of purified gap junction preparations was obtained from rodent liver by Hertzberg and Gilula (4) and Henderson et al. (5). Similar efforts led to the identification of a degradation fragment of about 29 kd of connexin43 from heart tissue, with subsequent isolation of the native connexin43 (7, 8). Efforts to isolate gap junctions from other tissues identified additional proteins which appear to be components of gap junctions or gap junction-like structures. Some of which are connexins (9, 11) and others (12, 13) which are not. The function of the latter groups of proteins remains controversial.

2.2 Isolation principles

Initial enrichment of liver gap junction-containing membranes is obtained by partial purification of bile canalicular plasma membranes which include gap junctions within their junctional domain. Non-junctional membrane is removed either subsequent to detergent solubilization or by sonication in the presence of alkali. Differential centrifugation further enriches for gap junctions which are then purified by density gradient centrifugation to yield connexin32 and connexin26.

2.3 Isolation of rat liver plasma membranes

Rat liver plasma membranes are isolated by homogenization in hypotonic buffer, differential centrifugation, and sucrose density gradient centrifugation. Especially when working with a large number of livers, it is best to work with two people, one sacrificing the animals and the other excising the livers. All steps are carried out with buffers on ice. In general, protease inhibitors need not be added to the buffers. If degradation of connexins in the final preparation is suspected, 1 mM phenylmethylsulfonyl fluoride (PMSF) or diisopropylfluorophosphate (DFP) can be added to all buffers until sample preparation for density gradient centrifugation. Frozen livers cannot be used.

The large-scale procedure is readily modified to work with smaller amounts of material, either with fewer rats or with mouse livers (50–100 mice).

Protocol 1. Large-scale isolation of rat liver plasma membranes

Solutions
- 1 mM $NaHCO_3$
- 37% sucrose (w/w), 1 mM $NaHCO_3$
- 55% sucrose (w/w), 1 mM $NaHCO_3$
- 67% sucrose (w/w), 1 mM $NaHCO_3$

Method
1. Sacrifice 40–50 female Sprague–Dawley rats (retired breeders preferred because of their relatively low cost) by guillotine after anaesthetizing in

Protocol 1. *Continued*

an atmosphere of carbon dioxide. This is easily established by placing dry ice in a styrofoam container and covering with several layers of newspaper. Excise livers and place in cold 1 mM $NaHCO_3$.

2. Trim fat and connective tissue from livers.
3. Mince liver into small pieces with scissors and rinse repeatedly with 1 mM $NaHCO_3$ until supernatant is clear. *Alternatively* tissue can be minced by a single passage through a Foley food mill. When the food mill is used, the minced tissue should NOT be rinsed because of loss of material from the more finely minced suspension.
4. Homogenize the minced tissue in an approx. tenfold excess of 1 mM $NaHCO_3$ in a loose-fitting Dounce homogenizer (4–8 strokes) monitoring breakage by phase-contrast microscopy. About 90–95% cell breakage is optimal. *Alternatively* the tissue can be homogenized using a Polytron at half-maximal power for 10–30 sec.
5. Dilute homogenate to about 4 litres with 1 mM $NaHCO_3$ and filter first through two layers and then four layers of cheesecloth.
6. Centrifugation to collect a crude membrane fraction can be carried out in a variety of ways depending upon availability of rotors. The most simple procedure involves the use of a continuous flow rotor (Beckman, JCF-Z) (a), although fixed angle rotors can be used (b).
    (a) Dilute homogenate further to about 7–8 litres with 1 mM $NaHCO_3$ and pump at 250 ml/min through JCF-Z rotor spinning at 3590 g.
    (b) Centrifuge in JA-10 rotor for 10 min at 4420 g with brake on low setting. Aspirate supernatant fraction and resuspend pellet in 1 mM $NaHCO_3$ with no more that two gentle strokes with a Dounce homogenizer. Repeat centrifugation 2–4 times until the supernatant fraction is relatively clear.
7. Resuspend pellet with gentle homogenization as in step 6(b) and centrifuge at 16 000 g for 10 min in order to reduce pellet volume prior to density gradient centrifugation.
8. Resuspend pellet in a minimal volume of 1 mM $NaHCO_3$ to give a final volume of about 250–300 ml. Measure this volume and add homogenate to two volumes 67% sucrose (w/w), 1 mM $NaHCO_3$. Filter through four layers of cheesecloth.
9. Density gradient centrifugation is used to remove nuclei and obtain a fraction enriched in bile canalicular plasma membranes. While the use of an ultracentrifuge zonal rotor is preferred (Beckman Ti-15) (a), angle rotors in a standard centrifuge can be used by prolonging the time of centrifugation (b).

(a) The zonal rotor can be loaded and unloaded at rest, greatly simplifying procedures in which loading and unloading are carried out while the rotor is rotating at low speed. A reorienting core can be used, but the procedure works adequately with a standard core. Place the sample in the four sectors of the rotor and then fill to top with 37% sucrose (w/w), 1 mM NaHCO$_3$. Adequate rates of flow with minimal disruption of the interface can be obtained using the loading cap which comes with the rotor. The rotor is then sealed and filled to capacity by introducing 55% sucrose (w/w), 1 mM NaHCO$_3$ with the loading cap (through the outer set of holes in the vanes). The use of a 60-ml syringe loaded with the sucrose solution permits its introduction under pressure until some 37% sucrose is forced out of the cap. The rotor is then placed in an ultracentrifuge and slowly accelerated to 84 000 g and maintained at this speed for an additional 1.5 h. The rotor is permitted to coast to a stop without the brake, immediately opened and the material at the sample/37% sucrose interface collected using a 50-ml syringe.

(b) The sample is divided into six bottles for a JA-10 rotor and overlayed with 37% sucrose (w/w), 1 mM NaHCO$_3$ to bottle capacity. Using slow acceleration, bring rotor to 16 000 g and continue centrifugation overnight (15–17 h). Let rotor coast to a stop without the brake and collect the material and the sample/37% sucrose interface.

10. The material is diluted two- to threefold with 1 mM NaHCO$_3$ and centrifuged in a JA10 rotor for 10 min at 16 000 g. The pellet is resuspended in a minimal volume of 1 mM NaHCO$_3$ and is stored at −20 °C until processed further for preparation of gap junctions.

2.3.1 Modifications for less tissue

The initial steps of the procedure are readily modified by centrifugation in rotors with smaller capacity, e.g. JA-14 and JA-20. Density gradient centrifugation can be carried out conveniently in bottles using a 35 rotor (Beckman) which can accommodate material from 7–10 rats or about 50 mice.

2.4 Isolation of rat liver gap junctions

Rat liver gap junctions prepared subsequent to treatment of an enriched plasma membrane fraction with either detergent or NaOH yields material with essentially the same polypeptide constituents, connexin32 and connexin26. While the two preparations are morphologically indistinguishable, circular dichroism studies have suggested that material prepared with the detergent Sarkosyl had somewhat higher content of α-helix than does that isolated after alkali extraction (14).

About 200–500 μg of purified gap junctions are obtained using the Sarkosyl

Isolation of intercellular junctions

procedure (*Protocol 2*) while as much as 2–3 mg subsequent to alkali treatment of plasma membranes (*Protocol 3*).

Protocol 2. Isolation of rat liver gap junctions using detergent

Solutions
- 1 mM NaHCO$_3$
- 0.1 M NaCl, 1 mM NaHCO$_3$
- 0.45% *N*-lauroyl sarcosine, Na salt (Sarkosyl NL-97), 1 mM NaHCO$_3$
- 0.1 M Na$_2$CO$_3$, pH 11
- 30% sucrose (w/w), 2 M urea, 1 mM NaHCO$_3$
- 30% sucrose (w/w), 1 M urea, 1 mM NaHCO$_3$
- 35% sucrose (w/w), 1 M urea, 1 mM NaHCO$_3$
- 45% sucrose (w/w), 1 M urea, 1 mM NaHCO$_3$
- 60% sucrose (w/w), 1 M urea, 1 mM NaHCO$_3$

Method
1. Dilute enriched plasma membranes (15–20 mg protein/ml) to about 80 ml with 1 mM NaHCO$_3$. Pellet by centrifugation at 17 400 g for 10 min in a JA-20 rotor.
2. Resuspend pellet to about 80 ml in 0.1 M NaCl, 1 mM NaHCO$_3$. Repeat the centrifugation.
3. Wash out excess salt by resuspending the pellet in 1 mM NaHCO$_3$ and repeating the centrifugation.
4. Resuspend pellet to 50 ml in 1 mM NaHCO$_3$ and divide into ten 5-ml aliquots. To each add 5 ml 0.45% *N*-lauroyl sarcosine, Na salt (Sarkosyl NL-97), 1 mM NaHCO$_3$.
5. Sonicate at setting 5 for 10 sec, using a sonicator (Model W-225R, Heat System Ultrasonics) with standard microtip. Foaming should be carefully avoided. Let stand at room temperature (RT) for 10 min and repeat the sonication. Let stand at RT for an additional 10 min.
6. Sediment by centrifugation at 43 700 g for 15 min in a JA-20 rotor (Beckman Instruments).
7. Aspirate off the supernatant fraction and resuspend the pellet in 40 ml of 1 mM NaHCO$_3$. Repeat the centrifugation.
8. Resuspend the pellet by sonication in 5 ml 0.1 M Na$_2$CO$_3$, pH 11. Repeat centrifugation and wash pellet in 5 ml 1 mM NaHCO$_3$.
9. Disperse pellet by sonication in 1.5 ml of 1 mM NaHCO$_3$. Add to this an equal volume of 30% sucrose (w/w), 2 M urea, 1 mM NaHCO$_3$.

10. Prepare discontinuous sucrose gradients in UltraClear tubes for the Beckman SW40 Ti rotor consisting of:
    (a) 0.5 ml of 60% sucrose (w/w), 1 M urea, 1 mM $NaHCO_3$.
    (b) 4 ml 45% sucrose (w/w), 1 M urea, 1 mM $NaHCO_3$.
    (c) 4 ml 35% sucrose (w/w), 1 M urea, 1 mM $NaHCO_3$.
    (d) 4 ml 30% sucrose (w/w), 1 M urea, 1 mM $NaHCO_3$.
    (e) 0.5 ml of the sample (15% sucrose, see step 8).
11. Carry out centrifugation in the SW40 Ti rotor at 270 000 g for 1.5 h.
12. Material is collected and pooled from interfaces using a 10-ml glass syringe with a 16- or 18-gauge cannula. It is easier to see the material at the interfaces by placing the tubes in front of a black piece of paper, and illuminate from the side. The material examined in this manner scatters light and appears white on a dark background. When all interfaces are collected, combine all of the remaining sucrose from the gradients as a 'gradient residue' fraction.
13. Dilute all fractions two- to threefold with 1 mM $NaHCO_3$. Collect by centrifugation at 43 700 g for 20 min in the JA-20 rotor. Resuspend by sonication in a minimal volume of 1 mM $NaHCO_3$, pool, dilute with 1 mM $NaHCO_3$ and repeat the centrifugation.
14. Resuspend the pellets by sonication is a minimal volume of 1 mM $NaHCO_3$. Purified gap junctions are obtained from the 35 to 45% sucrose interface; additional gap junctions are in the 'gradient residue' fraction.

Protocol 3. Isolation of rat liver gap junctions using alkali

Solutions
- 1 mM $NaHCO_3$
- 40 mM NaOH
- 20 mM NaOH
- 30% sucrose (w/w), 1 mM $NaHCO_3$
- 45% sucrose (w/w), 1 mM $NaHCO_3$
- 67% sucrose (w/w), 1 mM $NaHCO_3$

Method
1. Add 5 ml 40 mM NaOH to 5-ml aliquots of the enriched plasma membrane fraction (about 15–20 mg protein/ml) in polysulfone centrifuge tubes (polysulfone bottles tolerate alkali better than polycarbonate bottles).
2. Sonicate at setting 6 for 12 sec using a sonicator (Model W-225R, Heat System Ultrasonics) with standard microtip.

Isolation of intercellular junctions

Protocol 3. *Continued*

3. Add an additional 25 ml of 20 mM NaOH to each tube and mix.
4. Sediment by centrifugation at 43 700 g for 10 min in a JA-20 rotor (Beckman Instruments).
5. Aspirate off the supernatant fraction and loose material on top of the stratified pellet.
6. Resuspend pellet by adding about 250 μl 1 mM $NaHCO_3$ to each tube and disperse with gentle sonication. It is much easier to resuspend pellets if done in a small volume. Then dilute with addition buffer, pool material and repeat the 43 700 g centrifugation for 10 min.
7. Disperse pellet by sonication in a minimal volume of 1 mM $NaHCO_3$ then further dilute to about 12 ml. Add 16.54 g of 67% sucrose (w/w), 1 mM $NaHCO_3$, and adjust the volume to 25 ml. Confirm that the sample is 38% sucrose (w/w), adjusting if necessary. An inexpensive hand refractometer (Reichert-Jung 10431) is convenient for monitoring sucrose concentrations.
8. Prepare discontinuous sucrose gradients in UltraClear tubes for the Beckman SW40 Ti rotor by adding first 2 ml 45% sucrose (w/w), 1 mM $NaHCO_3$, followed by 4 ml of the sample in 38% sucrose, 5 ml 30% sucrose (w/w), 1 mM $NaHCO_3$, and then 1 mM $NaHCO_3$ to tube capacity.
9. Carry out centrifugation in the SW40 Ti rotor at 270 000 g for 1.5 h.
10. Material is collected and pooled from interfaces using a 10-ml glass syringe with a 16- or 18-gauge cannula. It is easier to see the material at the interfaces by placing the tubes in front of a black piece of paper, and illuminating from the side. The material examined in this manner scatters light and appears white on a dark background. When all interfaces are collected, combine all of the remaining sucrose from the gradients as a 'gradient residue' fraction.
11. Dilute all fractions two- to threefold with 1 mM $NaHCO_3$. Collect by centrifugation at 43 700 g for 20 min in the JA-20 rotor. Resuspend by sonication in a minimal volume of 1 mM $NaHCO_3$, pool, dilute with 1 mM $NaHCO_3$ and repeat the centrifugation.
12. Resuspend the pellets by sonication in a minimal volume of 1 mM $NaHCO_3$. Purified gap junctions are obtained from the 30 to 38% sucrose interface; additional gap junctions, often highly pure, are in the 'gradient residue' fraction.

2.4.1 Technical modifications for obtaining gap junctions from other tissues

A similar approach is also used for the isolation of myocardial gap junctions (7, 8, 15) and lens fibre junctions (16). While alkali treatment provides a residue further enriched in connexin43, this material does not provide suitable material for further purification of gap junctions.

2.5 Characterization of components of rodent liver gap junctions

Isolated rat liver gap junctions retain morphological features observed by electron microscopy of liver thin sections. These features include a pentalaminar or septilaminar structure in cross-section (*Figure 1B*) and a hexagonal array of particles with stain-penetrated spots in negative stain (*Figure 1D*). The protein constituents include a 27 kd protein and 21 kd proteins, along with other protein bands which are degradation products of the 27 kd protein (*Figure 1E*). The same protein constituents are present in isolated mouse liver gap junctions, although the 21 kd protein can account for as much as 40% of the total, compared to approx. 10% for rat liver gap junctions. The 27 kd protein and 21 kd protein are referred to as connexin32 and connexin26, respectively, based upon their predicted molecular masses from cDNA sequence analysis (17, 18). Lipid analysis of isolated rodent liver gap junctions has demonstrated a relative paucity of sphingomyelin (4) and enrichment for cholesterol (5).

2.6 Utility of preparation and future directions

Isolated liver gap junctions have been used for a number of types of studies, including ultrastructural analysis by X-ray diffraction and electron microscopy (19–24), reconstitution of channel activity (25, 26), topological analysis of the transmembrane distribution of specific segments of connexins (27–30), and identification of phosphorylated residues (31). It is anticipated that extension of these studies will continue to provide insight into the structure and function of gap junctions as well as the means by which these intercellular channels are regulated.

3. Tight junctions—Bruce R. Stevenson

3.1 Introduction

The tight junction (zonula occludens, occluding junction) plays a critical role in the normal functioning of epithelial and endothelial cells. It is the site of a selectively permeable barrier in the paracellular pathway and is believed to act as a fence restricting the movement of plasma membrane constituents between the compositionally distinct apical and basolateral domains (32–34). Despite its importance in epithelial cell biology, the tight junction is the intercellular junction about which the least is known biochemically (35, 36).

Isolation of purified subcellular fractions of tight junctions is, in theory, possible. Initial reports suggest that plasma membrane fractions derived from liver contain elements that morphologically resembled the tight junction

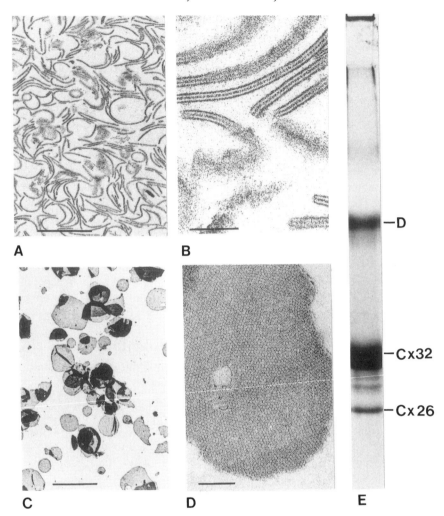

Figure 1. Characterization of isolated rat liver gap junctions. Isolated gap junctions were analysed by thin section (panels **A** and **B**) and negative stain electron microscopy (panels **C** and **D**) and by SDS-PAGE (panel **E**). Bar: 1 μm (panels **A** and **C**); 0.1 μm (panels **B** and **D**). The gap junctions retain their septillaminar appearance in thin section (panel **B**) and appear as arrays of subunits with a stain-delineated spot in the centre by negative stain (panel **D**). The positions of connexin32 (Cx32) and connexin26 (Cx26), as well as the dimer of connexin32 (**D**), on SDS-PAGE (panel **E**), are indicated. Their apparent molecular weights in this gel system are 27 kd, 21 kd, and 47 kd, respectively.

(37–39). Goodenough and Revel (40) further characterized the intercellular junctions in these fractions, and demonstrated that negative stain EM of liver plasma membranes briefly exposed to the detergent sodium deoxycholate (DOC) provided a useful morphological assay for the presence of tight junctions. These observations led to the development of an isolation protocol that enriched for liver junctional complexes, including the tight junction (41). The protocol relied on standard liver plasma membrane isolation methodologies followed by extensive exposure of a bile canalicular-enriched fraction to DOC in the presence of EGTA. A slightly modified version (ref. 42) of the original procedure is outlined below.

3.2 Isolation of a tight junction-enriched fraction from mouse liver

Protocol 4. Isolation of a tight junction-enriched fraction from mouse liver

Solutions (note: all solutions contain 0.2 mM DTT, 1 ml/litre aprotinin)
- BB: 1 mM sodium bicarbonate
- BB': 6 mM sodium bicarbonate
- HB: 1 mM sodium bicarbonate
  2.5 mg/litre leupeptin
  0.5 mg/litre pepstatin
  0.5 mg/litre chymostatin
  30 mg/litre soybean trypsin inhibitor
- Sucrose solutions contain:
  1 mM sodium bicarbonate pH 8
  2 mM EGTA
- M: 10 mM imidazole pH 8
  1 mM EGTA
- MD: 0.5% DOC
  10 mM imidazole pH 8
  1 mM EGTA

Methods
Day 1
1. Sacrifice 100 mice, five at a time, by cervical dislocation. Mice 4–6 weeks of age are optimal. Livers from older animals result in higher levels of connective tissue contamination; younger animals have smaller livers and give correspondingly smaller yields. All further operations and solutions, unless noted, are performed at 4°C.
2. Homogenize livers five at a time in solution BB with 10 strokes in a 40-ml Dounce homogenizer (loose pestle) submerged in ice.
3. Dilute homogenate immediately into a 3 litre solution BB'.
4. Allow total diluted homogenate (about 4 litres) to stand at 4°C for 10 min; filter through 32 layers of cheesecloth.

Isolation of intercellular junctions

Protocol 4. *Continued*

5. Spin homogenate 10 min at 2750 g with **no brake**. Aspirate off supernatant, including any white material (fat) floating on surface. Handle centrifuge bottles gently during aspiration; be careful to avoid aspirating any of the very loose pellet.
6. Resuspend pellets by vigorously swirling in BB. Combine and dilute to about 1 litre with BB, and spin as in step 5. Aspirate off supernatant as before. At no point in the protocol should a pellet be resuspended by pipetting. The bile canaliculi and detergent extracted junctional ribbons are sensitive to shear.
7. Resuspend pellet(s) in 154 ml BB. Add 196 ml 67% (wt/wt) sucrose slowly with stirring. Pour a two-step gradient: 41% (sample) overlaid with 25% (wt/wt) sucrose in BB. Spin 90 min at 150 000 g. Collect thick pad at interface. A 12 ultracentrifuge tube gradient [e.g. Beckman SW28] is more convenient than a zonal rotor gradient and gives a cleaner interface. Under phase microscopy this interface should consist of bile canaliculi with adjacent membrane sheets with minimal small vesicular contamination. This material is usually allowed to sit overnight in sucrose partly because it's convenient and partly because it leads to better detergent extractions.

Day 2

8. Dilute the material collected in step 7 to approx. 140 ml with solution M. Spin in four 50-ml tubes, 15 min at 25 000 g. Discard supernatant.
9. Wash pellets with M, spinning 10 min at 3600 g in four 50-ml tubes. Discard supernatant. Wash membranes with M until the supernatants are relatively clear of small vesicular material. The pellets from step 8 are very hard and may be difficult to resuspend. Gentle trituration with a wide-mouth pipette can be used if necessary. Use a vortex set at a medium speed for all subsequent resuspensions.
10. Bring centrifuge, rotor, and solution M to RT. The elevated temperature is critical for optimal detergent extractions.
11. Resuspend each of the four pellets in 10 ml M. Add slowly while vortexing 10 ml 1% DOC in M. Spin 10 min at 12 000 g. Discard supernatant. Do not add the protease inhibitor PMSF to any detergent extraction solution. This lipophilic compound inhibits membrane solubilization.
12. Resuspend pellets in about 70 ml solution MD by vortexing. Spin in two 50-ml tubes, 10 min at 1500 g. Save the supernatant. Repeat wash of pellets.
13. Resuspend pellets in one 50-ml tube and spin as in step 11. Save the supernatant. Discard pellet. (While the pellet contains numerous junctional ribbons, it also contains a large amount of collagen and other insoluble detritus.)

14. Combine supernatants saved in steps 11 and 12 and spin in four 50-ml tubes, 10 min, 12 000 g. Save the pellets, discard supernatant.

15. Resuspend the pellets in approximately 35 ml MD. Allow any remaining large insoluble particles visible to the eye to settle to the bottom of the tube and transfer the supernatant to another 50-ml tube. Spin 10 min at 13 000 g. Discard supernatant. The final pellet is the DOC-extracted junctional ribbon fraction [DOC-JR]. High-power phase microscopy of a this pellet resuspended in 1–2 ml of MD reveals a population of elongated worm-like remnants of the junctional complex. This fraction is stable in DOC. Yield should be on the order of a 2–4 mg total protein from 100 mice.

It should be emphasized that *Protocol 4* results in a relatively impure preparation, especially in comparison to the strategies employed for isolating the other intercellular junctions described in this chapter. SDS-PAGE analysis of the DOC-JR fraction reveals numerous silver-stained bands, and evaluation by thin section or negative stain EM (*Figure 2*) provides evidence for copious non-tight junctional contamination, including gap junctions and elements of the zonula adherens. However, this preparation has aided in the identification of at least one tight junctional component (42).

3.3 Modifications, variations, and future directions

Liver is used as starting material because it is a readily available, it is rich in tight junctions, and several protocols have been developed for the isolation of liver plasma membranes. Hypothetically, plasma membrane fractions from other epithelial tissues containing tight junctions could also be used. Preliminary work suggests that tight junctions from chicken intestinal epithelial cell brush borders are resistant to extraction with Triton X-100 (43). However, fractions obtained from this source were heavily contaminated with brush border cytoskeletal proteins.

Although mouse liver is employed in the procedure outlined here, rat liver can also be used. In a slightly modified protocol, the methodologies of Hubbard *et al.* (44) are used to generate bile canaliculus-enriched plasma membrane fractions. Prior to detergent extraction membranes (in solution M) are brought to pH 11 with NaOH for one hour. In addition, the DOC-JR fraction from rat liver appears less stable in detergent; hence, junctional ribbons are pelleted immediately after exposure to DOC and washed once in solution M (L. A. Jesaitis and D. A. Goodenough, personal communication).

It would obviously be desirable to enrich this preparation further for elements of the tight junction. However, the tight junction is a relatively fragile plasma membrane specialization which shows little resistance to purification procedures that have been used on other intercellular. Selective detergent solubilization and mechanical disruption, which facilitate the

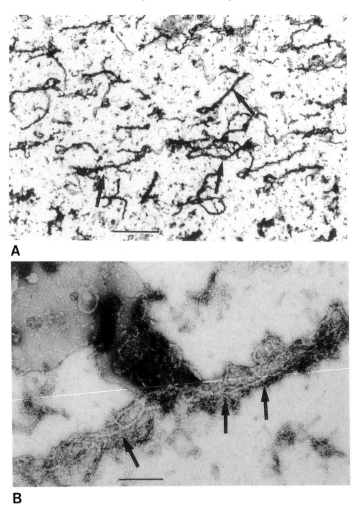

Figure 2. Negative stain electron micrographs of deoxycholate-extracted junctional ribbons (DOC-JR) derived from mouse liver. (**A**) Representative low magnification view in which numerous elongate junctional ribbons can be seen (*arrows*). Smaller, non-junctional contamination is also visible. Bar: 3 μm. (**B**) High magnification of one junctional ribbon reveals branching and anastomosing fibrils (*arrows*) embedded within the ribbon matrix. Immediately juxtaposed to the junctional ribbon is a gap junction. Bar: 0.2 μm.

isolation of desmosomes and gap junctions, have not been successfully applied. For example, treatment of DOC-JR with the detergent N-lauroyl–sarcosine leaves the junctional ribbons intact, but fibrils contained within these ribbons (*Figure 2*), believed to correspond to the strands seen in freeze-fractured tight junctions, are no longer visible in negative stain EM (41). It is not known whether fibril components are solubilized by *N*-lauroyl-sarcosine or whether their structural arrangement within the membrane is altered so as to preclude visualization.

Mechanical disruption treatments (for example, sonication) completely disrupt junctional ribbon structure and, like most other detergent treatments, leave behind a significant population of the intact gap junctions.

Despite its relative lack of purity, the mouse liver-derived DOC-JR fraction was used to generate monoclonal antibodies against ZO-1, a high-molecular-weight phosphoprotein peripherally associated with the cytoplasmic surface of tight junction membranes in epithelial and endothelial cells (42, 45). In addition, a 160 kd peripheral membrane polypeptide which co-immunoprecipitates with ZO-1 has recently been described. This protein, tentatively named ZO-2, may form a complex with ZO-1 at the tight junction (46). Cingulin, a third peripheral membrane protein found at the tight junction, was fortuitously discovered during the production of monoclonal antibodies against chicken brush border myosin (47). Although the identification of these peripheral membrane proteins indicates an intricate system operating on the cytoplasmic surface of the junctional membrane, no information is currently available on the putative transmembrane component(s) of the tight junction. However, use of known junctional proteins as markers in the modification and extension of isolation strategies provided in this section may yield further knowledge on the molecular constituency of the tight junction.

4. Desmosomes—Kathleen J. Green

4.1 Introduction

Desmosomes are intercellular junctions that play dual roles in cell adhesion and in providing specific cell surface attachment sites for intermediate filaments (IF) (for reviews, see refs 48–51). It is thought that the IF-desmosome network may redistribute the forces created by mechanical stress during embryogenesis and in adult tissues. Desmosomes are abundant in epithelial tissues, but are also found in myocardial and Purkinje fibre cells of the heart, the arachnoidal cells of the meninges, and the dendritic reticulum cells of germinal centres of lymph nodes.

At the ultrastructural level, desmosomes are characterized by a pair of plasma membranes separated by an intercellular space of approx. 30 nm that is bisected by a line, called the central dense stratum, and connected by cross-striations (50–52). Proximal to the plasma membranes lie extremely electron dense plaques approximately 15–17 nm in width (*Figure 3*, 'ODP' or outer dense plaque). A thinner, relatively electron lucent layer is sandwiched between the outer dense plaque and an amorphous fuzzy layer (*Figure 3*, 'IDP' or inner dense plaque) through which IF or tonofilaments appear to loop. Together these layers of the cytoplasmic plaque can extend up to 40 or 50 nm

Isolation of intercellular junctions

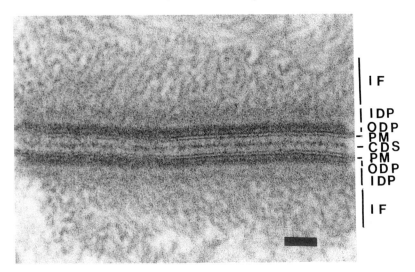

Figure 3. Electron micrograph of a single bovine tongue desmosome. PM = plasma membrane, CDS = central dense stratum, ODP = outer dense plaque, IDP = inner dense plaque, IF = intermediate filaments. Bar: 50 nm.

on either side of the plasma membrane. Classification of a junction as a desmosome is based not only on structural criteria, but also on the presence of specific protein and glycoprotein markers, whose identification has been made possible by the ability to obtain large quantities of enriched preparations of these junctions.

Skerrow and Matoltsy (53) published the first account of desmosome isolation in 1974. They utilized selective extraction in low pH buffer followed by discontinuous sucrose density gradient centrifugation to isolate desmosomes from bovine muzzle. The stratified epithelia of bovine muzzle and tongue have been popular sources of desmosomes due to the high density of junctions in these tissues, and easy access to large quantities from local slaughterhouses. Several modifications of the original procedure have been reported, including approaches for further fractionation into the adhesive (core) and cytoskeleton-attachment (plaque) domains. With only minor modifications, however, the original procedure still stands as the most straightforward approach for isolating enriched fractions of desmosomes largely separated from attached keratin filaments.

4.2 Isolation principles

Extraction of cellular constituents from epidermis typically requires harsh conditions of extreme pH or denaturing buffers, reflecting the protective biological function of skin. For isolation of desmosomes, conditions used previously to extract epidermal tissues, NaOH and urea, were rejected since the former dissolved desmosomes and the latter extracted the desmosomal plaque. The buffer chosen instead was citric acid–sodium citrate (pH 2.6) which disrupts cell membranes and selectively extracts epidermal keratins, leaving desmosomes, for the most part, intact (54). Sucrose density gradient centrifugation is then used to separate desmosomes from other cellular components. During this procedure, nuclei, dermal fragments and cell debris all pellet to the bottom. Lighter membrane fragments and vesicles band higher. Most laboratories currently use a modification of the original citrate buffer isolation procedure, one of which is presented here. This modification employs the use of sonication and detergent to disrupt interdesmosomal membranes and to free desmosomes from intracellular organelles, and the addition of protease inhibitors, both of which were recommended by Gorbsky and Steinberg (55). Other minor changes were reported by Jones *et al.* (56).

4.3 Isolation of enriched preparations of desmosomes from bovine tongue

Protocol 5. Isolation of enriched preparations of desmosomes from bovine tongue

Materials

Tissue
Two unpigmented cow tongues from a local slaughterhouse (freshly slaughtered and kept on ice). (About 30 g wet weight of stripped epidermis per tongue yields about 50–60 mg of purified desmosomes.)

Solutions

- tongue buffer   6 mM Na^+/K^+ phosphate, pH 7.0
                         120 mM NaCl
                         3 mM KCl
                         20 mM Na_2EDTA

   Just prior to use, add the following to the tongue buffer to a final concentration of:

                         1 mM PMSF (stock solution in EtOH)
                         2 mM DTT
                         5 μm/ml pepstatin (stock in MeOH)
                         5 μg/ml leupeptin (stock in H_2O)
                         2 mM TAME (add solid)

- citric acid/sodium citrate   0.1 M citric acid (free acid) pH to 2.6 with 0.1 M
  (CASC buffer)   sodium citrate (or use NaOH)
- CASC/Triton   CASC plus 0.1% Triton X-100 or NP-40, plus
                               cocktail of protease inhibitors as above.

Isolation of intercellular junctions

Protocol 5. *Continued*

- CASC/sucrose   40, 50, 55, and 65% sucrose solutions (w/v) in 0.1 M CASC
- PBS (phosphate-buffered saline)   8 M urea (deionized)
- Urea sample buffer   1% SDS
  10% glycerol
  60 mM Tris–HCl (6.8)
  0.003% pyronin Y
  5% BME added just prior to use

Methods

1. Remove the epidermis and underlying connective tissue from the tongue in four lengthwise strips of about 1–2 cm using a sharp razor blade. Scrape off all underlying muscle, disturbing the dermis/epidermis as little as possible.
2. Collect these strips in a beaker containing 500 ml tongue buffer with freshly added protease inhibitors. After the dissection is complete, change the buffer and incubate overnight at 4°C.
3. Strip the epidermis away from the dermis using rat-tooth forceps to grip the dermis and regular forceps to grip the epidermis. Discard the dermis. Mince the epidermal tissue finely with scissors.
4. Add 80 ml of CASC/Triton solution and stir at 4°C for 2 h.
5. Sonicate (with stirring at 4°C) for 10 treatments, 15 sec each, with 10 sec rest intervals using a large probe. (Heat Systems W-220F sonicator, power setting 7, or Branson Sonifier Cell Disruptor 185, power setting 8 or 9.)
6. Filter the homogenate through four layers of cheesecloth.
7. Divide the filtrate between the two 40-ml tubes, and centrifuge the filtrate at 10 000 g in a Sorvall (RC-5B) SS34 rotor for 20 min. Pipette off supernatant being careful not to disturb the loose pellet.
8. Resuspend the pellet by vortexing in 20 ml per tongue of CASC buffer containing 1 mM PMSF. [*Note*: The desmosome-enriched portion of the pellet will resuspend easily, possibly leaving a discoloured tighter portion of the pellet. Do not attempt to completely disaggregate this portion as it is depleted in desmosomes and will spin to the bottom of the sucrose gradient (see below).]
9. Repeat this washing procedure at least two times. (Each wash serves to extract additional keratin protein.) After the final spin, resuspend pellets in 10 ml per tongue of 40% sucrose in 0.1 M CASC.
10. Layer a discontinuous sucrose gradient in each of two 40-ml tubes (for SW28 Beckman centrifuge rotor) with 9.5 ml each of 60, 55, and 50%

sucrose in 0.1 M CASC. Store the solutions at 4°C and allow gradient to sit for 15 min at 4°C prior to use.

11. Carefully overlay the 40% sucrose solution containing desmosomes over the gradient. Spin in an SW28 rotor at 58 400 g for 3 h at 4°C without the brake.

12. Remove the 55–60 (best) and 50–55 interfaces from each tube with a Pasteur pipette. Pool each interface and dilute up to about 40 ml with PBS–2 mM PMSF or 0.1 M CASC. Pellet in an SS34 rotor in the Sorvall for 30 min at 12 000 g at 4°C. (*Note*: If one desires to resuspend desmosomes without sonication or mechanical disruption after storage as a pellet, do these washes and store in CASC. If neutral pH is desired, e.g. for EM analysis, do washes and store in PBS. PBS will cause aggregation of desmosomal material.)

13. Store as a frozen pellet or resuspend in urea sample buffer for subsequent SDS-PAGE analysis.

4.4 Technical modifications of standard protocol

(a) The order and number of centrifugation and sonication steps prior to sucrose gradient centrifugation can be varied. In the procedure described in *Protocol 5*, sonication is done prior to initial centrifugation as described by Jones *et al.* (56); however, it can be carried out on the pelleted material following an initial centrifugation (55, 57). In addition, the use of a Polytron or similar homogenizer can be used prior to an initial centrifugation if desired (53). A note of caution regarding homogenization or sonication: overdoing these steps can cause disruption of the desmosome proper, resulting in anomalous migration in the sucrose gradient, and a depletion of desmosomes from the 55–60% interface. To increase the yield of desmosomes, long term stirring at 4°C in CASC/Triton (step 4) is preferable to over-homogenization.

(b) Differential centrifugation can be used as an alternative to the sucrose gradient step (55). After sonication and low speed centrifugation (750 g for 20 min) to remove residual cells and nuclei, a higher speed spin at 12 000 g for 20 min results in a trilaminar pellet with desmosomes at the top. After washes, this portion of the pellet is of sufficient purity for many applications.

(c) Drochmans *et al.* (58), reported an alternative procedure which involves swelling the epidermis in hypotonic, moderately alkaline solution (pH 9–11) and mechanical disruption. The resulting plasma membranes with attached desmosomes were further fractionated by sonication, followed by discontinuous sucrose density centrifugation. Although this method yielded enriched fractions of desmosomes, keratin-containing tonofilaments remain attached, and thus this procedure is not used as widely as the standard Skerrow and Matoltsy preparation.

Isolation of intercellular junctions

4.5 Further fractionation of bovine tongue desmosomes

Protocol 6. Further fractionation of bovine tongue desmosomes

1. Metrizamide gradients were utilized by Gorbsky and Steinberg to purify the glycoprotein components of desmosomes (55). After rinsing and further sonication of enriched desmosomes obtained from the trilaminar pellet described above, desmosomes were centrifuged through linear 10–45% gradients of metrizamide as described in detail in ref. 55. It was suggested that the metrizamide interacts selectively with the protein components of the desmosomes, releasing them from the membrane-bound glycoproteins.
2. Denaturing agents such as urea and guanidinium hydrochloride can be used to further fractionate desmosomes following the standard protocol. Plaque components are selectively extracted in 9 M urea, pH 7.5 or pH 9.0, whereas the transmembrane core components are insoluble under these conditions (56, 57). [It should be noted, however, that contrary to this, Franke *et al.* (59) have reported the selective partial extraction of two transmembrane core components under similar conditions.]
3. Addition of 6 M guanidinium HCl in the CASC during initial cell lysis selectively extracts plaque proteins, enriching for transmembrane core glycoproteins (60). Hydroxyapatite column chromatography in the presence of SDS can then be used to fractionate core components.

4.6 Characterization of desmosomal components

A typical Skerrow and Matoltsy preparation comprises about 76% protein, 17% carbohydrate, and 10% lipid (61). The major non-keratin bands include four plaque proteins, dp1–4 and four core glycoproteins dg1–3 shown in *Figure 4C*. The nomenclature used here, 'dp' (desmosomal protein); and 'dg' (desmosomal glycoprotein), was agreed upon by the major labs in the field at the 1987 CIBA Meetings (51); however, previous terminology also remains in widespread use.

Six of the major components, dp1,2,3 and dg1,2,3 are thought to be constitutively expressed in all desmosomes studied to date (50). The highly related proteins, dp1 and 2 (otherwise known as desmoplakins I and II or DPI and II) migrate at about 240 and 210 kd by SDS-PAGE, although we now know this to be a significant underestimate from analysis of cDNA clones. These are the major plaque components, comprising about 35% by weight of desmosomal proteins, and have been shown by immunoelectron microscopy to be localized in the innermost, amorphous region of the plaque (IDP). Dp1 and 2 have now been purified to homogeneity using extensive extraction in 4 M urea followed by ion exchange and gel filtration chromatography (62).

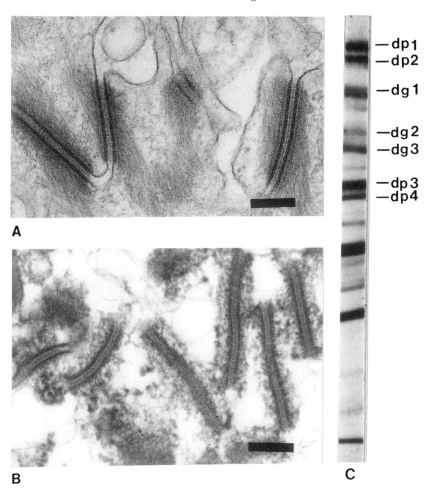

Figure 4. Isolation of bovine tongue desmosomes. (**A**) Electron micrograph of bovine tongue desmosomes *in situ*. (**B**) Electron micrograph of isolated desmosomes. Bars: 0.25 μm. (**C**) 7.5% SDS-PAGE gel of preparation seen in (**B**), with the major desmosomal proteins and glycoproteins identified: dp = desmosomal protein, dg = desmosomal glycoprotein. Apparent molecular weights are 240 kd (dp1), 210 kd (dp2), 160–165 kd (dg1), 110 kd (dg2), 100 kd (dg3), 81 kd (dp3), and 75 kd (dp4).

Biochemical and morphological analysis (62), as well as predictions based on cDNA structure (reviewed in ref. 50), indicate that dp1 preferentially forms an extended dumbbell-shaped dimer with a central rod domain about 132 nm in length. Dp3, also known as plakoglobin, is an approx. 81 kd protein that is unlike other major desmosomal proteins in that it is also found in microfilament-associated adherens junctions. Plakoglobin is a protein of 7S with a Stokes radius of 5 nm and a molecular mass suggestive of a dimer (50).

The glycoproteins dg1 at about 165 kd [(otherwise known as desmoglein I (DGI), or band 3)] and dg2/3 at 110 and 100 kd [(otherwise known as desmocollins I and II, desmogleins II a and b (DGIIa,b), or band 4,a,b] are thought to be constitutive desmosomal components, although the possibility of tissue-specific forms of dg2/3 has been left open (50). As the major constituents of the 'desmoglea' or intercellular 'glue' these are candidate cell adhesion molecules (48–52). Recent sequence data demonstrating homology to the calcium-dependent cadherin class of cell adhesion molecules is consistent with this notion. Other prominent bands ranging from about 40 to 65 kd are thought to represent five bovine keratin polypeptides.

In addition, to these major components, a number of other constituents of the desmosome have been reported, using fractions similar to those described here as starting material. Some of these have a restricted distribution and may be relatively minor in abundance (reviewed in ref. 50). Dp4, although a major (approx.) 75 kd plaque component in bovine tongue and muzzle preparations (*Figure 4C*) has been reported to be present in stratified epithelial tissues only. Several investigators have reported varying amounts of a 22 kd glycoprotein, dg4 (desmoglein III), in bovine muzzle desmosome preparations. The calmodulin-binding protein desmocalmin, comigrating with dp1 at about 240 kd (perhaps the same as the so-called 'D1' antigen), has been reported in bovine muzzle and several other stratified tissues, but it is not yet clear how widespread the distribution of this component is. An extremely high molecular mass protein of about 680 kd, named desmoyokin, is absent from desmosomes of simple epithelia and myocardium. A minor 140 kd component antigenically related to nuclear lamin B, and the (approx.) 300 kd IFAP known as plectin have also been localized immunologically to desmosomal plaques (49). Plectin's cellular localization is not restricted to desmosomes, however. This molecule is thought to act as a major linker for a variety of cytoskeletal components. Other proteins reported to be localized, but not restricted to, desmosomal membrane domains are E-cadherin (uvomorulin) and the 140 kd pemphigus vulgaris antigen (50).

The precise functions of these components is for the most part unknown. With few exceptions, purification of these desmosomal constituents in a native state has been hampered by the insolubility of the desmosome. For the same reasons, determination of protein–protein interactions has been difficult. Several components, largely found in the plaque, have been implicated in linking IF to the cell surface; these include dp1/2, the 140 kd lamin-like protein, the 75 kd protein, and desmocalmin. Dp1/2 is the most abundant, and only confirmed constitutive component, in this group. However, *in vitro* evidence indicating direct binding of dp1/2 to IF is lacking. With regard to the core proteins, direct evidence from antibody inhibition assays has demonstrated a role for dg2/3 in cell adhesion; however, a function for dg1 in adhesion has not yet been demonstrated.

4.7 Future directions

Skerrow and Matoltsy noted the existence of at least 24 bands in their desmosome preparations; several have yet to be characterized, and additional components not visible on their gels have been identified. Of the major components that have been characterized extensively, surprisingly little is known of their function. This may be due in part to the difficulty in isolating proteins in their native state to study protein–protein interactions using traditional biochemical approaches. The recent isolation and characterization of cDNA clones for dp1,2,3 and dg1,2,3 will now facilitate studies of a functional nature using a molecular genetic approach.

With novel desmosomal proteins being identified regularly, it is clear that the number that are necessary and sufficient to form a desmosome has not yet been established. Fractions such as those described here are likely to continue to provide a source of novel desmosomal constituents; however, it is equally likely that important components are extracted by the acidic buffer conditions used in the procedure. Thus, in the future, alternative approaches for isolating potentially soluble proteins will need to be determined.

5. Cell-to-cell adherens junctions—Shoichiro Tsukita

5.1 Introduction

The cell-to-cell adherens junctions (AJ), also referred to as belt desmosomes, are defined as one of the typical cell–cell contacts in which actin filaments are associated with the plasma membrane through its well-developed undercoat (63, 64). The adhesion molecules concentrated at the cell-to-cell AJ were recently shown to be cadherins [E-cadherin/uvomorulin, N-cadherin/A-CAM, P-cadherin, and L-CAM] (65). The cell-to-cell AJ is commonly believed to be directly involved in tissue morphogenesis. This feature is, most likely, related to the tight association between AJ and the force-generating actin filament bundles. In addition to these mechanical properties of AJ, the notion has become widespread that AJ plays crucial roles in the control of cell growth and tissue morphogenesis through transmembrane signalling via cadherin molecules.

Various types of tissues bear the typical cell-to-cell AJ. In most AJ-bearing cells such as intestinal epithelial cells, liver cells, and retinal pigment cells, the AJ shows a belt-like appearance, while in cardiac muscle cells the patch-like AJ is well developed at the intercalated discs. Recently, using rat liver cells, we succeeded in developing an isolation procedure for the belt-like AJ (66). Early isolation attempts were made in many labs using cardiac muscle cells or retinal pigment cells (67, 68). However, both preparations were not appropriate for analysing the molecular architecture of the cell-to-cell AJ proper, mainly because in both types of cells the actin filament bundles associated with AJ were highly developed.

5.2 Isolation principles

In the liver, hepatic cells are arranged to form laminae, one to two cells thick, and the apposed sides of neighbouring cells form a bile canaliculus between them. The principal and essential steps of our newly developed procedure for the isolation of the cell-to-cell AJ from rat liver are summarized in *Figure 5*. This isolation procedure includes two steps. As the first step, the fraction rich in bile canaliculi ('B.C. fraction') was isolated according to a modification of the method developed by Song *et al.* (39). In the second step, the fraction rich in the long belts of AJ ('AJ fraction') was recovered from the B.C. fraction by treatment with Nonidet P-40 (NP-40).

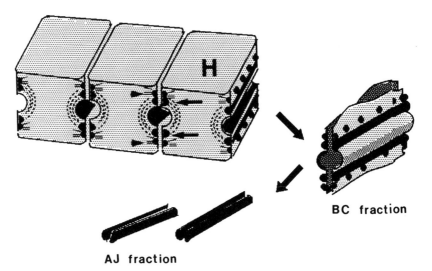

Figure 5. A scheme for the isolation of adherens junctions from rat liver. H = hepatic cells; arrowheads = desmosomes; arrows = adherens junctions.

5.3 Isolation of adherens junctions from rat liver

Protocol 7. Isolation of adherens junctions from rat liver

Solutions
- 1 mM $NaHCO_3$, 2 mg/ml leupeptin, pH 7.5
- 55% (w/v) sucrose
- 42.9% (w/v) sucrose
- 100 mM KCl, 1 mM $MgCl_2$, 0.1% (vol./vol.) NP-40, 10 mM Hepes, pH 7.5
- 20% (w/v) sucrose

- 50% (w/v) sucrose
- 1 mM EGTA, 0.5 mM PMSF, 2 mM Tris–HCl, pH 9.2
- 1 M acetic acid, pH 2.3

Method

1. Six 8-week-old Wistar rats are decapitated, the livers carefully taken out and soaked in an ice-chilled physiological saline solution. All subsequent procedures were carried out at 4°C.
2. The liver is minced with razor blades and treated with 'hypotonic solution' consisting of 1 mM $NaHCO_3$ and 2 mg/ml leupeptin, pH 7.5, for 30 min. The swollen samples are homogenized in 2 vol. of hypotonic solution by the use of a loose-fitting Dounce homogenizer.
3. The homogenate is diluted with hypotonic solution to 900 ml, filtered twice through four layers of gauze, and centrifuged at 1500 g for 10 min.
4. The pellets are resuspended in hypotonic solution, diluted to 26.2 ml, and mixed with 193.8 ml of 55% (w/v) sucrose solution to make a 48.5% sucrose solution.
5. The sample is divided into six tubes, carefully layered with 42.9% sucrose solution and centrifuged for 60 min at 100 000 g. The bile canaliculi were recovered at 42.9:48.5% interface.
6. The fraction rich in bile canaliculi is diluted with 10 vol. of hypotonic solution and centrifuged for 30 min at 4000 g.
7. The pellet obtained is resuspended in NP-40 solution consisting of 100 mM KCl, 1 mM $MgCl_2$, 0.1% (vol./vol.) NP-40, 10 mM Hepes, pH 7.5. The sample is passed twice through a 23-gauge needle and stirred in NP-40 solution for 20 min. It is then fractionated by centrifugation at 100 000 g for 60 min on a discontinuous gradient consisting of 20% (w/v) and 50% (wt/vol.) sucrose. Adherens junctions were recovered at the 20:50% interface (*Figures 6* and *7*).

To analyse the molecular architecture of the isolated AJ, we have dissected the isolated AJ as follows. The isolated AJ was dialysed against the low-salt extraction solution (1 mM EGTA, 0.5 mM PMSF, and 2 mM Tris–HCl [pH 9.2]) at 4°C overnight, followed by centrifugation at 100 000 g for 1 h. By repeating the low-salt extraction three times, most of the undercoat-constitutive proteins were extracted as the 'AJ extract', leaving the 'AJ membrane' fraction. This AJ membrane still includes some tightly-bound undercoat-constitutive proteins, which are selectively extracted with 1 M acetic acid (pH 2.3) for 30 min at 4°C as the 'acid extract', leaving the 'acid membrane'. Therefore, to analyse the molecular organization of the undercoat of AJ, the AJ extract and the acid extract were used, while for the analysis of the integral membrane proteins the acid membrane was used.

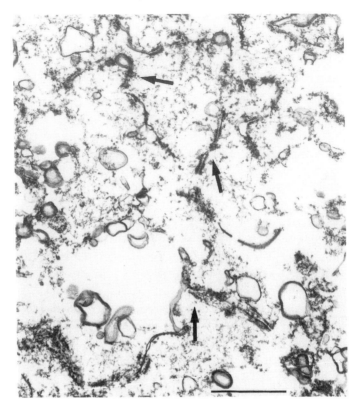

Figure 6. Thin-section electron microscopy of the fraction rich in adherens junctions (*arrows*). Bar: 1 μm.

5.4 Technical modifications

This isolation procedure is applicable not only for rat but also for mouse, bovine, and probably other mammals. This isolation procedure does not work well when chick liver is used as a starting material. To obtain the plasma membrane proper from the AJ membrane, 0.1 M NaOH, or 3 M guanidine can be substituted for 1 M acetic acid.

5.5 Characterization of components of the isolated AJ

Since no biochemical enrichment procedure has been available for AJ before now, immunological approaches were the only way to identify components of the undercoat of this type of junction. Vinculin (130 kd) was the first protein to be reported to be localized exclusively in the cell-to-cell and cell-to-substrate AJ (69). α-actinin (100 kd) was found in both types of AJ, but this protein was also localized in the stress fibres of cultured non-muscle cells (70). Some other structural proteins have been shown to be localized in the cell-to-

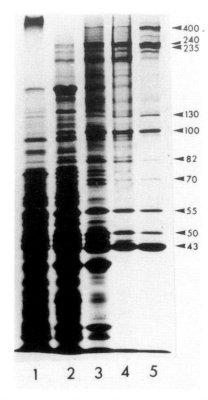

Figure 7. SDS-PAGE of fractions obtained during isolation of adherens junctions. *Lane 1*, rat liver homogenate; *lane 2*, pellet after homogenization of liver with hypotonic solution and centrifugation; *lane 3*, fraction rich in bile canaliculi; *lane 4*, fraction rich in adherens junction; *lane 5*, AJ extract. The molecular mass of specific proteins are indicated (kd).

cell AJ by immunological methods: plakoglobin (71), one of the major components of desmosomal plaque, and an 82 kd protein (72) identified using an autoimmune serum.

The development of the isolation procedure for the cell-to-cell AJ from liver cells has prompted us to systematically analyse the molecular organization of the cell-to-cell AJ. As described above, the isolated AJ can be dissected into the undercoat ('AJ extract' and 'acid extract') and the plasma membrane proper ('acid membrane').

5.5.1 Major components of AJ extract

AJ extract is mainly composed of ten polypeptides with apparent molecular masses of 400, 240, 235, 130, 100, 82, 70, 55, 50, and 43 kd. It was shown biochemically and immunologically that 240 kd, 235 kd, 130 kd, 100 kd, and 43 kd polypeptides were α-spectrin, β-spectrin, vinculin, α-actinin, and actin, respectively. So far, two of the remaining five polypeptides, 400 kd and 82 kd proteins, were purified and characterized.

[Radixin]
An 82-kd protein ('radixin') has been purified from AJ extract (73). The substoichiometric radixin largely inhibited the actin filament assembly; when the molar ratio of radixin to G-actin was 1:1000, the viscosity was reduced to 28% of the control value. Direct electron microscopic studies revealed that radixin selectively inhibited monomer addition at the barbed ends of actin filaments. By use of the antibody raised against radixin, this protein was shown by immunofluorescence microscopy to be localized at the cell-to-cell AJ in various types of cells. In contrast, radixin was not concentrated at the cell-to-substrate AJ. Therefore, it was concluded that the 82 kd protein is a barbed end-capping protein which is associated with the undercoat of the cell-to-cell AJ. Most recently, it was clearly shown that in various types of cells radixin was highly concentrated at the cleavage furrow during cytokinesis (74). The most prominent feature shared by AJ and the cleavage furrow is the tight association of actin filaments with plasma membranes, suggesting that radixin plays an important role in the association of actin filaments with the plasma membrane in general.

[Tenuin]
Monoclonal antibodies specific for the 400 kd polypeptide ('tenuin') of AJ extract were obtained (75). Immune blot analyses showed that tenuin occurs in various types of tissues. Immunofluorescence microscopy and immunoelectron microscopy have revealed that tenuin is distributed not only at the undercoat of AJ but also actin bundles associated with AJ in non-muscle cells: stress fibres in cultured fibroblasts and circumferential bundles in epithelial cells. The partially purified protein molecule looks like a slender rod about 400 nm in length. Therefore, tenuin may play a crucial role in forming and maintaining the actin-filament-based cytoskeletal network, including actin bundles and AJ in non-muscle cells.

5.5.2 Major components of acid extract
Acid extract is composed of 5–6 polypeptides. The most major polypeptide with an apparent molecular mass of 102 kd has been shown to be identical to one of the cadherin-associated proteins (CAP 102 or α-catenin), which were reported to be co-immunoprecipitated with cadherin molecules by anti-cadherin antibody (76). This protein was purified from the acid extract and the cDNA encoding CAP 102 was isolated. Sequence analysis of the cDNA revealed that CAP 102 is similar to vinculin.

5.5.3 Major components of acid membrane
10 polypeptides can be reproducibly identified in the acid membrane. Almost

all of them appear to be glycosylated. The 130 kd band was shown to be identical to E-cadherin.

5.5.4 Others

By *in vitro* phosphorylation experiments with isolated AJ and immunoblotting analyses, it was recently shown that two specific proto-oncogenic tyrosine kinases of *src* family (*c-yes* and *c-src*) are highly enriched in the cell-to-cell AJ where the level of tyrosine phosphorylation is elevated (77).

5.6 Summary

It is of central importance in this field to understand at a molecular level how the undercoat of AJ regulates cell–cell adhesion and how it participates in signal transduction from adhesion molecules to the nucleus. After the detailed cataloguing of the major constituents of the undercoat of AJ using isolated AJ, just as was true in the study of erythrocyte membranes, molecular analysis of their associations will help clarify the key roles of AJ in the regulation of cell–cell adhesion and in signal transduction.

Acknowledgements

The authors wish to thank Dr Rattan Nath and Laura Nilles for critical reading of the desmosome portion of the chapter.

E.L.H. is supported by a grant from the NIH (GM30667) and is a recipient of an Irma T. Hirschl Career Scientist Award. K.J.G. is supported by grants from the American Cancer Society (#BE-56 and Junior Faculty Research Award), The Council for Tobacco Research USA, Inc. #2432, NIH (HD24430), and the March of Dimes (Basil O'Connor Starter Scholar Research Award). S.T. is supported by grants from the Ministry of Education, Science and Culture, Japan. B.R.S. is supported by grants from the Medical Research Council of Canada, and the Alberta Heritage Foundation for Medical Research.

References

1. Barr, L., Dewey, M. M., and Berger, W. (1965). *J. Gen. Physiol.*, **48**, 797.
2. Barr, L., Berger, W., and Dewey, M. M. (1968). *J. Gen. Physiol.*, **51**, 347.
3. Robertson, J. D. (1963). *J. Cell Biol.*, **19**, 201.
4. Hertzberg, E. L. and Gilula, N. B. (1979). *J. Biol. Chem.*, **254**, 2138.
5. Henderson, D., Eibl, H., and Weber, K. (1979). *J. Mol. Biol.*, **132**, 193.
6. Nicholson, B., Dermietzel, R., Teplow, D., Traub, O., Willecke, K., and Revel, J. P. (1987). *Nature (Lond.)*, **329**, 732.
7. Manjunath, C. K., Goings, G. E., and Page, E. (1982). *Biochem. J.*, **205**, 189.
8. Manjunath, C. K., Goings, G. E., and Page, E. (1985). *J. Membr. Biol.*, **85**, 159.

9. Manjunath, C. K., Nicholson, B. J., Teplow, D., Hood, L., Page, E., and Revel, J. P. (1987). *Biochem. Biophys. Res. Commun.*, **142**, 228.
10. Benedetti, E. L. and Emmelot, P. (1968). *J. Cell Biol.*, **38**, 15.
11. Kistler, J., Christie, D., and Bullivant, S. (1988). *Nature (Lond.)*, **331**, 721.
12. Johnson, R. G., Klukas, K. A., Tze-Hong, L., and Spray, D. C. (1988). In *Gap Junctions* (ed. E. L. Hertzberg, and R. G. Johnson), pp. 81–98. Alan R. Liss, New York.
13. Gruijters, W. T., Kistler, J., Bullivant, S., and Goodenough, D. A. (1987). *J. Cell Biol.*, **104**, 565.
14. Cascio, M., Gogol, E., and Wallace, B. A. (1990). *J. Biol. Chem.*, **265**, 2358.
15. Kensler, R. W. and Goodenough, D. A. (1980). *J. Cell Biol.*, **86**, 755.
16. Goodenough, D. A. (1979). *Invest. Ophthalmol. Vis. Sci.*, **18**, 1104.
17. Paul, D. L. (1986). *J. Cell Biol.*, **103**, 123.
18. Zhang, J. T. and Nicholson, B. J. (1989). *J. Cell Biol.*, **109**, 3391.
19. Makowski, L., Caspar, D. L., Phillips, W. C., and Goodenough, D. A. (1977). *J. Cell Biol.*, **74**, 629.
20. Caspar, D. L., Goodenough, D. A., Makowski, L., and Phillips, W. C. (1977). *J. Cell Biol.*, **74**, 605.
21. Makowski, L. (1988). *Adv. Cell Biol.*, **2**, 119.
22. Zampighi, G. and Unwin, P. N. (1979). *J. Mol. Biol.*, **135**, 451.
23. Gogol, E. and Unwin, N. (1988). *Biophys. J.*, **54**, 105.
24. Sosinsky, G. E., Baker, T. S., Caspar, D. L., and Goodenough, D. A. (1990). *Biophys. J.*, **58**, 1213.
25. Young, J. D., Cohn, Z. A., and Gilula, N. B. (1987). *Cell*, **48**, 733.
26. Spray, D. C., Saez, J. C., Brosius, D., Bennett, M. V., and Hertzberg, E. L. (1986). *Proc. Natl Acad. Sci. USA*, **83**, 5494.
27. Zimmer, D. B., Green, C. R., Evans, W. H., and Gilula, N. B. (1987). *J. Biol. Chem.*, **262**, 7751.
28. Hertzberg, E. L., Disher, R. M., Tiller, A. A., Zhou, Y., and Cook, R. G. (1988). *J. Biol. Chem.*, **263**, 19105.
29. Milks, L. C., Kumar, N. M., Houghten, R., Unwin, N., and Gilula, N. B. (1988). *EMBO J.*, **7**, 2967.
30. Goodenough, D. A., Paul, D. L., and Jesaitis, L. (1988). *J. Cell Biol.*, **107**, 1817.
31. Saez, J. C., Nairn, A. C., Czernik, A. J., Spray, D. C., Hertzberg, E. L., Greengard, P., and Bennett, M. V. (1990). *Eur. J. Biochem.*, **192**, 263.
32. Gumbiner, B. (1987). *Am. J. Physiol.*, **253**, C749.
33. Gumbiner, B. (1990). *Curr. Opin. Cell Biol.*, **2**, 881.
34. Stevenson, B. R., Anderson, J. M., and Bullivant, S. (1988). *Mol. Cell. Biochem.*, **83**, 129.
35. Stevenson, B. R. and Paul, D. L. (1989). *Curr. Opin. Cell Biol.*, **1**, 884.
36. Anderson, J. M. and Stevenson, B. R. (1992). In *Tight Junctions* (ed. M. Cereijido). CRC Press.
37. Neville, D. M. (1960). *Cytology*, **8**, 413.
38. Emmelot, P., Bos, C. J., Benedetti, E. L., and Rumke, P. H. (1964). *Biochim. Biophys. Acta*, **90**, 126.
39. Song, C. S., Rubin, W., Rifkind, A. B., and Kappas, A. (1969). *J. Cell Biol.*, **41**, 124.

40. Goodenough, D. A. and Revel, J. P. (1970). *J. Cell Biol.*, **45**, 272.
41. Stevenson, B. R. and Goodenough, D. A. (1984). *J. Cell Biol.*, **98**, 1209.
42. Stevenson, B. R., Siliciano, J. D., Mooseker, M. S., and Goodenough, D. A. (1986). *J. Cell Biol.*, **103**, 755 (Abstract).
43. Stevenson, B. R., Goodenough, D. A., and Mooseker, M. S. (1982). *J. Cell Biol.*, **95**, 92a (Abstract).
44. Hubbard, A. L., Wall, D. A., and Ma, A. (1983). *J. Cell Biol.*, **96**, 217.
45. Anderson, J. M., Stevenson, B. R., Jesaitis, L. A., Goodenough, D. A., and Mooseker, M. S. (1988). *J. Cell Biol.*, **106**, 1141.
46. Gumbiner, B., Lowenkopf, T., and Apatira, D. (1991). *Proc. Natl Acad. Sci. USA*, **88**, 3460.
47. Citi, S., Sabanay, H., Jakes, R., Geiger, B., and Kendrick Jones, J. (1988). *Nature (Lond.)*, **333**, 272.
48. Garrod, D. R., Parrish, E. P., Mattey, D. L., Marston, J. E., Measures, H. R., and Vilela, M. J. (1990). In *Morphoregulatory molecules* (ed. G. M. Edelman, B. A. Cunningham, and J. P. Thiery), pp. 315–39. John Wiley, New York.
49. Jones, J. C. R. and Green, K. J. (1991). *Curr. Opin. Cell Biol.*, **3**, 126.
50. Schwarz, M. A., Owaribe, K., Kartenbeck, J., and Franke, W. W. (1990). *Ann. Rev. Cell Biol.*, **6**, 461.
51. Block, G. and Clark, S. (ed.) (1987). *Junctional Complexes of Epithelial Cells*. John Wiley, Chichester.
52. Steinberg, M. S., Shida, H., Giudice, G. J., Shida, M., Patel, N. H., and Blaschuk, O. W. (1987). In *Junctional Complexes of Epithelial Cells* (ed. G. Bock and S. Clark), pp. 3–25. John Wiley, Chichester.
53. Skerrow, C. J. and Matoltsy, A. G. (1974). *J. Cell Biol.*, **63**, 515.
54. Matoltsy, A. G. (1965). In *Biology of the Skin and Hair Growth* (ed. A. G. Lyne and B. F. Short), pp. 291–305. American Elsevier, New York.
55. Gorbsky, G. and Steinberg, M. S. (1981). *J. Cell Biol.*, **90**, 243.
56. Jones, S. M., Jones, J. C. R., and Goldman, R. D. (1987). *J. Cell Biochem.*, **36**, 223.
57. Skerrow, C. J., Hunter, I., and Skerrow, D. (1987). *J. Cell Sci.*, **87**, 411.
58. Drochmans, P., Freudenstein, C., Wanson, J.-C., Laurent, L., Keenan, T. W., Stadler, J., Leloup, R., and Franke, W. W. (1978). *J. Cell Biol.*, **79**, 427.
59. Franke, W. W., Kapprell, H.-P., and Mueller, H. (1983). *Eur. J. Cell Biol.*, **32**, 117.
60. Blaschuk, O. W., Manteuffel, R. L., and Steinberg, M. S. (1986). *Biochem. Biophys. Acta*, **883**, 426.
61. Skerrow, C. J. and Mtoltsy, A. G. (1974). *J. Cell Biol.*, **63**, 524.
62. O'Keefe, E. J., Erickson, H. P., and Bennett, V. (1989). *J. Biol. Chem.*, **264**, 8310.
63. Geiger, B. (1983). *Biochim. Biophys. Acta*, **737**, 305.
64. Tsukita, Sh., Tsukita, Sa., and Nagafuchi, A. (1990). *Cell Struct. Funct.*, **15**, 7.
65. Takeichi, M. (1988). *Development*, **102**, 639.
66. Tsukita, Sh. and Tsukita, Sa. (1989). *J. Cell Biol.*, **108**, 31.
67. Colaco, C. A. and Evans, W. H. (1982). *Biochim. Biophys. Acta*, **684**, 40.
68. Owaribe, K. and Masuda, H. (1982). *J. Cell Biol.*, **95**, 310.
69. Geiger, B. (1979). *Cell*, **18**, 193.
70. Lazarides, E. and Burridge, K. (1975). *Cell*, **6**, 289.

71. Cowin, P., Kapprell, H., Franke, W. W., Tamkun, J., and Hynes, R. O. (1986). *Cell*, **46**, 1063.
72. Beckerle, M. C. (1986). *J. Cell Biol.*, **103**, 1679.
73. Tsukita, Sa., Hieda, Y., and Tsukita, Sh. (1989). *J. Cell Biol.*, **108**, 2369.
74. Sato, N., Yonemura, M., Obinata, M., Tsukita, Sa., and Tsukita, Sh. (1991). *J. Cell Biol.*, **113**, 321.
75. Tsukita, Sh., Itoh, M., and Tsukita, Sa. (1989). *J. Cell Biol.*, **109**, 2905.
76. Nagafuchi, A., Takeichi, M., and Tsukita, Sh. (1991). *Cell*, **65**, 849.
77. Tsukita, Sa., Oishi, K., Akiyama, T., Yamanashi, Y., Yamamoto, T., and Tsukita, Sh. (1991). *J. Cell Biol.*, **113**, 867.

7
The *Xenopus* oocyte cell–cell channel assay for functional analysis of gap junction proteins

GERHARD DAHL

1. Introduction

The paired *Xenopus* oocyte expression system was originally designed as a functional assay to identify mRNAs involved in the formation of intercellular channels, the functional units of the gap junction (4). Subsequently, the paired oocyte assay was used to demonstrate that the expression of a single exogenous gene, rat connexin32, could induce intercellular communication (3). Thus, a functional criterion for the identity of a gap junction or cell–cell channel-forming protein was established. More recently, a number of genes encoding proteins highly related to connexin32 have been cloned (1). This proliferation of potential gap junction proteins has increased the utility of functional expression systems designed to examine cell–cell channels. The experimental capabilities of such systems are as follows:

- determination of the intercellular channel forming capability of cloned proteins and detailed comparison of their properties
- expression cloning (2) of intercellular channel forming proteins not related to connexins or from organisms where cloning procedures based on DNA homology to previously identified connexins have been unsuccessful
- structure–function analysis of intercellular channel proteins by *in vitro* mutagenesis

Two classes of expression systems can serve these purposes: (*i*) mRNA or DNA injected *Xenopus* oocytes (3, 4), and (*ii*) transfected cell lines (5, 6). In this chapter, we will describe the use of the paired *Xenopus* oocyte expression system which offers the following advantages:

- Speed: Assuming the DNAs to be tested are available in an appropriate vector, they can be tested for the ability to produce cell–cell channels in 1–3 days. In contrast, expression systems involving cultured cells require the

establishment of stabile transfected lines which can take several weeks or months.
- Low equipment costs: Since oocytes are very large cells, dissecting microscopes are adequate for all aspects of the procedure. Inexpensive coarse micromanipulators may be used to carry injection and recording microelectrodes and since relatively low resistance electrodes are required, sophisticated microelectrode pullers are not necessary.
- Quantitation and statistical reliability: It is feasible to test a large number (10–100) oocyte pairs in a few hours.
- The endogenous cell–cell channel activity can be avoided or suppressed by several methods, which will be outlined later in the chapter.

Although the paired *Xenopus* oocyte system offers significant advantages in some applications, it is not suitable for the study of single cell–cell channel currents. A dual whole-cell patch-clamp system for the analysis of single gap junction channel currents between cultured cells is described in Chapter 8.

2. The paired Xenopus oocyte expression assay

The oocyte expression system pioneered by Gordon and co-workers (7) has long been the preferred vehicle for exogenous expression because these large and robust cells are easy to handle and to inject. Furthermore, their low metabolic activity and their large spare translational capacity (8) makes them ideal for expression studies. While many oocytes exhibit such features, those from *Xenopus laevis* are particularly suited because they do not exhibit the pricking response (maturation into eggs upon pricking of membrane; for example, during injection) that many other oocytes do.

The paired *Xenopus* oocyte expression assay consists of the following steps:

(a) Surgical removal of ovaries from HCG primed frogs.

(b) Separation of follicular cells from Stage VI oocytes.

(c) Preparation and injection of RNA.

(d) Manual removal of the vitelline envelope.

(e) Assembly of oocyte pairs.

(f) Impalement with microelectrodes for dual-cell voltage-clamp.

2.1 Frogs: handling and injection of HCG

Xenopus laevis, the African clawed frog, has its natural habitat in South Africa. These animals are easy to rear and are now commercially available from farms in many countries. In the US, guaranteed oocyte-bearing frogs may be purchased from Xenopus I or Nasco. Frogs may be kept (1 animal in

3–4 litres of water) in large tanks filled to a depth of 10–15 cm. Heavy lids should be fitted as the frogs can jump up to several centimetres above the water level and also climb up in corners. Tap water is appropriate provided it is dechlorinated by storing the water in a holding tank for 1 to 2 days and bubbling it with air through an air stone. The water temperature should be between 18–22°C. Temperature transients in excess of 5°C should be avoided at all times. The water should be changed twice to three times a week, after feedings. Feeding can be done twice a week with frog brittle and once a week with chopped beef or calf liver. To avoid fouling the water, spoon-feeding of the liver is advisable.

The adult female has a body length of over 8 cm, and sex identification is easily done on the basis of the cloacal lips that are present only in the female. Furthermore, a female with full-grown ovaries characteristically shows bulging flanks. The ovaries typically contain oocytes of all maturation stages (9). Stage V and stage VI oocytes, which are usually used for expression studies, amount to only about 10%. Upon injection of human chorion gonadotropin (HCG; Sigma Chemical Co.) into the dorsal lymph sac, the frog will shed the fully mature oocytes and a new generation of oocytes will progress to maturity. 500 IU of HCG should be injected about 1 month before harvesting oocytes. This results in partial synchronization of the ovary and a higher yield of oocytes useful for expression studies. Surgical removal of ovarian lobules reportedly has a similar effect (10). Xenopus I and Nasco provide the service of HCG-injection and will ship female frogs at specified times after injection.

2.2 Preparation of oocytes

To prepare oocytes, the whole ovaries may be removed after sacrificing the frog. Alternatively, one may surgically remove portions of the ovary (ovarian lobes) and use the same donor several times. Note that even with fully standardized procedures there is variability in the response of oocytes from the same ovary and even more so between oocytes from different ovaries. To obtain statistically reliable data one should aim, therefore, for more than 10 oocytes pairs for each experimental condition. Furthermore, each experiment should include negative and positive controls; for example, pairs of uninjected oocyte and pairs injected with a wild-type connexin mRNA, respectively. With a little safety margin for oocyte loss during the incubation period, therefore, more than 100 oocytes are needed to test one experimental condition.

After removal from the frog the oocytes are dissociated and defolliculated in one step with collagenase treatment, following *Protocol 1*. More discussion of the surgical techniques are provided in another volume in this series (see Alan Colman's chapter in *Transcription and Translation: A Practical Approach*; ed. B. D. Hames and S. J. Higgins).

The Xenopus oocyte cell–cell channel assay

Protocol 1. Obtain and defolliculate oocytes

1. Anaesthetize the frog by placing under ice for about 20 min. Animal may then be sacrificed or kept on ice for the remainder of the surgery if re-use is desired.
2. Make small cut (1–2 cm) through the skin and body wall off the midline of the abdomen. The ovary, which almost completely fills the abdomen, can be pulled out with tweezers.
3. After excising the lobe, or lobules, the incision may be sutured closed if the frog is to be re-used. Usually, two stitches each in the body wall and the skin are sufficient.
4. Cut open individual lobules with scissors.
5. Prepare 0.2% collagenase in 'Ca-free' OR2 (see below) in suspension culture flask (slow-speed stirring vessel).
6. Cut ovary into small pieces and drop into collagenase solution.
7. Stir with magnetic stirrer at low r.p.m. (about 1/sec) for 2–3 h at room temperature (RT).
8. Check visually for oocytes to separate.
9. Wash exhaustively with OR2.
10. Inspect under stereo-microscope and sort oocytes according to stage of maturation and physical appearance.
11. Collect good oocytes. These should have an even pigmentation and be devoid of all follicle cells but have an intact vitelline layer.[a]
12. Store oocytes in tissue culture dishes with OR2 medium (with Ca^{2+}) at about 19°C until injection.

Composition of incubation medium OR2 (12):
'Ca^{2+} free' OR2 is prepared without $CaCl_2$

NaCl	82.5 mM
KCl	2.5 mM
$CaCl_2$	1.0 mM
$MgCl_2$	1.0 mM
Na_2HPO_4	1.0 mM
Hepes	5.0 mM

polyvinylpyrrolidone 0.5 g/litre pH 7.5

for longer storage of oocytes antibiotics should be included:

Penicillin	50 000 U/µl
Streptomycin	50 mg/µl

[a] Oocytes have to be handled carefully from this stage on. For safe handling it is recommended to make all transfers by pipetting using, for example, a 200-µl glass pipette (such as Acupette) attached to a mouthpiece. The pipette, like all other glassware, should be fire-polished over a bunsen burner to avoid damaging the oocytes on sharp edges.

2.3 Synthesis of mRNA

The most efficient functional expression of cell–cell channels in paired oocytes is obtained with injection of *in vitro* synthesized connexin-specific mRNA. However, the assay is sufficiently sensitive to yield detectable amounts of channels from unpurified mRNA extracted from tissues actively expressing gap junctions (4). A variety of mRNA extraction protocols have been developed, and the reader is referred to standard texts for details.

Gene expression can also be effected by injection of DNA into the nucleus of the oocyte. In this case, the gene of interest is carried in a eukaryotic expression vector containing DNA elements capable of directing transcription, termination, and/or polyadenylation. Functional expression of connexin32 in *X. laevis* oocytes has been achieved in this manner, employing a heat shock promoter (3). However, the low overall survival rate of oocytes after heat shock renders this strategy much less efficient than direct injection of RNA.

For preparation of synthetic connexin-specific mRNA the connexin gene is subcloned into a suitable plasmid containing a bacteriophage RNA polymerase promoter, e.g. SP6, T7, T3. The plasmid is linearized by cutting the DNA downstream of the insert. Linearization may be performed at any site within the vector except the promoter and protein coding regions. Restriction enzymes generating 3' overhangs should not be used for linearization since this may result in undesired transcription of the antisense strand. Addition of the appropriate RNA polymerase (see *Protocol 2*) yields RNA transcripts of defined sequence and length.

Because translational efficiency of mRNA is affected by 5' untranslated sequences (11) and by the 3' tail (12, 13) it is important that the 5' and 3' extensions of the coding sequences be comparable, if not identical, for similar expression levels to be obtained. Connexin genes are reconstructed so that a Kozak consensus sequence precedes the ATG codon. The 3' tail is represented by approx. 400–600 bases of vector sequence and is generated during the linearization step. Shorter tails yield transient responses, suggesting that RNA stability is reduced.

Protocol 2. *In vitro* transcription

Materials
- RNase-free water
- SP6 polymerase (20 000 U/ml)[a]
- acetylated BSA (10 mg/ml)
- RNasin (38 000 U/ml) (Promega-Biotech)
- 10 mM ATP, CTP, GTP, UTP (Boehringer Mannheim) pH 7.0, sterile-filtered and stored in aliquots at $<-20\,°C$

Protocol 2. *Continued*
- 1 mM GTP
- 5 mM GpppG (Pharmacia)
- 100 mM DTT
- 5 × SP6 buffer: 200 mM Tris–HCl pH 7.5, 30 mM $MgCl_2$, 10 mM spermidine, 50 mM NaCl
  - 2 μg/μl linearized plasmid DNA containing connexin sequence

Procedure
1. Mix in the following proportions, for example:

water	4
SP6 buffer	4
DTT	2
BSA	1
RNasin	1.4
ATP	1
CTP	1
UTP	1
GTP (1 mM)	1
GpppG	2

2. Add linearized template DNA: 2 μl.

   This is a convenient branch point if more than one mRNA preparation is to be made, multiples of the above mixture are divided into aliquots before adding the various DNA samples.
3. Add SP6 polymerase: 2 μl.
4. Incubate at 40°C: for 60 min.
5. Add GTP (10 mM): 1 μl.
6. Incubate at 40°C: for 30 min.
7. Check RNA production (yield, proper size, etc.) by gel electrophoresis. (See *Figure 1*.) Standard protocols ask for DNase treatment and phenol/chloroform extraction of RNA before loading on to agarose gel. It is, however, possible to load an aliquot of the whole transcription reaction. The proteins do not seem to interfere with DNA and RNA migration in the gel, as indicated by their electrophoretic mobility in relation to appropriate markers. This procedure makes it possible in one step to ascertain whether the template was properly linearized, the synthesized mRNA is of appropriate size, and the RNA yield was adequate. The RNA band should be more intense than the DNA band if the transcription reaction was efficient.

[a] The protocol, a modification of the original procedure (14), uses SP6 polymerase. Specifications for other polymerases differ slightly.

Figure 1. Agarose gel electrophoresis of transcription reaction mixture. *Right lane*: The upper band represents the linearized template DNA, the middle band the synthetic mRNA, and the lower band is a segment of vector DNA generated during linearization. The vector is pGEM-3Z f(+) with connexin43 inserted between the *Bam*H1 and *Eco*R1 sites of the multiple cloning region. Linearization for transcription was performed using SSP I. Since there are three 3 SSP I sites, restriction generates fragments of ~3.9 kb, 0.4 kb, and 24 bp. The smallest fragment is not visible in this gel. Transcription using SP6 RNA polymerase produced an RNA of 1.84 kb. *Left lane*: DNA size markers of 23.1, 9.4, 6.6, 4.3, 2.3, and 2.0 kb.

A common cause for a low yield of mRNA (provided polymerase activity is adequate) is a limiting concentration of nucleotides; in particular, the 1 mM GTP is critical.

2.4 Purification of RNA

Ordinarily, *in vitro* synthesized mRNA is purified from the transcription mixture before oocyte injections. This is usually accomplished by DNase treatment, followed by phenol–chloroform, and one or more ethanol precipitations. These steps are lengthy and therefore increase the risk of RNA loss and introduction of RNases. In addition, ethanol precipitated RNA may assume secondary structures that may interfere with efficient translation. Alternatively, one can inject the entire transcription reaction mixture after the incubation scheme is completed 'as is' into the oocytes. In our experience this turned out to be the most efficient and reliable procedure.

When kept refrigerated the transcript remains fully active for more than a week. The only disadvantage is that the quantitation of RNA is somewhat more cumbersome. It can be done with sufficient accuracy by gel electrophoresis as shown in *Figure 1*, including dilutions and appropriate standards.

2.5 Microinjection into oocytes

The costs for injection apparatus range from US$63 to several thousand dollars for a computer-controlled pneumatic device. The US$63 variety consists of a micrometre syringe (Gilmont) connected to a glass capillary via rigid Teflon tubing. The syringe and tubing should be filled with hydraulic fluid for dampening; brake fluid or extra virgin olive oil work well. However, a small air buffer should be present to avoid contamination of the mRNA solution with hydraulic fluid. The injection needles are prepared from glass capillaries (1 mm outer diameter, 0.8 mm inner diameter) on a microelectrode puller. Almost any puller capable of producing low resistance electrode will be appropriate. We have used vertical pullers from Narashige (Model PE-2) and Kopf instruments (Model 730). With appropriate settings of the heating element and the solenoid, tip diameters between 1–10 μm can be obtained. This is the range of tip openings suitable for oocyte injections. Alternatively, higher resistance electrodes can be broken under a compound microscope to yield appropriately sized openings. In case visual control of the injection process is desired, an elongated thin shaft should be produced. For this purpose vertical pullers are the easiest to use. With a smaller diameter, the distance that the fluid meniscus moves during injection will increase enough to be readily observed. Such a shaft is made by initially pulling the glass to a stop located a few millimetres below the starting point. A second free pull then generates the actual tip.

It is useful to have a working place equipped with a stereo-microscope, proper illumination (fibre glass ring optics), and a micromanipulator. This set-up can be used for oocyte dissecting and injections. A variety of low-power, coarse micromanipulators can be used to hold the injection pipette. Many investigators have used the Singer pantographic type manipulator (Oxford Microinstruments) which has the advantage of speed at the expense of stability. It requires practice to become skilled. We have successfully used simple three-axis manipulators (World Precision Instr. M3301), clamped (WPI M5 clamp) to a rod projecting from a magnetic base (WPI MI base). For the ultimate stability, we use a manipulator composed of modules available from Newport Instruments. Two translation stages (model 430) are bolted on top of a magnetic base (model 150) at 90-degree angles to each other to provide X and Y axis movement. A 45-degree base (model 360–45) is added with a third translation stage fixed to it. This provides movement along an axis ideal for oocyte penetration.

Protocol 3. mRNA injection

1. Pipette about 2 μl of mRNA solution on parafilm-covered support.
2. Manipulate the needle tip into droplet.
3. Turn micrometer to generate negative pressure.

4. After the needle is filled to desired level, turn the micrometre to generate slight positive pressure, in order to avoid dilution and contamination of RNA solution with medium.
5. Retract needle and replace support for mRNA with culture dish containing oocytes to be injected in OR2 medium.
6. Advance needle into medium and adjust flow of RNA solution. If total transcription reaction is used its glycerol content comes in handy here because of the difference in optical density to the medium.
7. With tweezers in one hand, manipulate oocyte one by one into the path of the needle, leaving the tweezers behind the oocyte for mechanical support while the needle is advanced into the oocyte rapidly with the coarse control knob of the micromanipulator.
8. Wait until desired volume (about 20–50 nl) is dispensed (see *Protocol 4*: Calibration), retract the needle and move the new oocyte into place (keeping track of injected oocytes!)
9. Discard any damaged oocytes, transfer good ones into fresh OR2 medium and incubate at 19°C.

The beginner is advised to practise these steps with a dye solution; for example, phenol red. In a successful injection the oocyte will turn light pink in the neighbourhood of the injection site, and the colour fades out at a distance of 200–300 μm

---

Protocol 4. Calibration of injection volume

1. Put 2 markers (e.g. score with diamond) on needle as indicated in schematic.
2. Fill needle with solution containing molecules that can easily be quantitated (e.g. any radioactive isotope).
3. Eject 5 vol. between two marks on filter papers.
4. Pipette 5 μl of same solution on another filter paper for reference.
5. Dry filters, add scintillant, and count.
6. Calculate volume.

2.6 Removal of the vitelline envelope

At this stage of the procedure the oocytes are still covered with a layer of a glycoprotein meshwork that is several microns thick. Although microvilli invade this vitelline layer it is very rare that microvilli of two opposing oocytes could meet. For establishment of intimate membrane contact, a prerequisite of cell–cell channel formation, it is therefore necessary to remove the vitelline layer.

The Xenopus oocyte cell–cell channel assay

In the absence of commercially available specific enzymes that would break down vitelline this has to be done by hand. Modified Dumont No. 5 'Biology' tweezers can be used for this purpose. The modification involves thinning of the tips, blunt ending the tips with a slight angle, and roughing of the inner surface of the tip to get a good holding grip. The modification is done with ultrafine emery paper under a dissection microscope.

For stripping the vitelline layer the open forceps are advanced onto the oocyte surface and then gently closed. If the forceps gets a grip of the vitelline layer they are slightly pulled back so that a second pair of forceps gets a hold of the vitelline layer right next to it. When both forceps have a tight grip they are pulled apart gently. This will rupture the vitelline envelope which then can easily be peeled of the rest of the oocyte. This procedure requires steady hands. It is important that basic ergonomic rules are followed in setting up the dissection stand, which includes firm support for arms and hands.

It is much easier to grab the vitelline layer and remove it when the oocytes have been shrunk in hypertonic medium. Under this condition the vitelline layer is detached from the oocyte surface and the forceps can get a hold of it with little danger of stabbing the oocyte. This procedure is popular among researchers performing patch-clamp studies (15). Because these are acute experiments where the recording from isolated membrane patches starts within minutes after removal of the vitelline layer, artefacts due to the hypertonic solution are unlikely to be encountered. For the cell–cell channel assay, on the other hand, one has to consider serious artefacts stemming from this procedure. It usually takes several hours until junctional conductance is determined after removal of the vitelline layer. During this time both translational and post-translational events, including transport of connexin proteins to the plasma membrane, take place. Such events could easily be affected by change of the ionic strength in the cytoplasm, a possible consequence of hypertonic shock. Furthermore, the oocyte is loaded with maternal RNA—including RNA for cell–cell channel proteins to be made in the early embryo. Most of the maternal RNA is not biologically active, because of sequestration by specific binding proteins (12). It is conceivable that the binding of protein to maternal RNA is affected by excessive changes in ionic strength of the cytoplasm. Indeed, in comparison to regularly stripped oocytes the ones stripped after hypertonic shock tend to exhibit a higher level of endogenous channel activity (E. Levine and G. Dahl, unpublished). In addition to lower levels of exogenous expression, we observe instable membrane potentials and lower long-term survival rates in oocytes subjected to hypertonic shock.

Beginners in the trade of oocyte stripping may use an aid. One can slightly lift the vitelline layer off the oocyte membrane by placing a fire-polished small glass capillary on to the surface, applying gentle suction and pulling the capillary carefully back. At the edge, usually a piece of vitelline layer is lifted sufficiently to get a hold of it easily with forceps. Once this hold is established, the capillary is put aside and a second forceps is used to tear the vitelline layer for removal as described above.

From this point on the oocytes have an unprotected naked membrane and are extremely fragile. In addition they tend to stick to the support. It is essential to take precautionary measures as listed below:

- perform all transfers of oocyte by mouth pipetting with fire-polished pipettes
- all media should be filtered and checked for absence of fibres that could stab oocytes
- oocytes must never get to the air–water interface, including gas bubbles
- media should be de-gassed under vacuum
- use non-sticky surfaces for oocyte storage, i.e. Teflon, parafilm, suspension culture dishes
- if oocytes are to be treated before pairing with enzymes, site-specific reagents, peptides, antibodies, etc. (16), sticking to the support surface can be prevented by gentle shaking either by hand or with an orbital shaker at low setting

With these precautions in effect oocytes survive incubation periods of several hours without harm.

2.7 Doping and pairing

It will take 4–6 h until functional cell–cell channels can first be detected, if the pairing is performed shortly after mRNA injection. If pairing follows mRNA injection with a delay period (more than 12 h) the channels are detected earlier and the rate of formation increases (17, 18). There is, however, considerable variability in response to RNA between different oocyte preparations. This variability can be reduced by doping the oocytes with lectins. Glycine max, triticum vulgaris, and concanavalin A are equally effective. In addition, application of lectins before pairing results on the average in ten times higher junctional conductances than seen in oocytes pairs without treatment. Oocytes should be incubated in 10 µg/ml lectin in OR2 for 20 min. Lectin should then be removed with a brief OR2 wash. Removal of steric hindrance seems to be the main mechanism responsible for the lectin effect on cell–cell channel formation (19).

Pairing of oocytes is conveniently done in Teflon grooves, with the oocytes held in place by stoppers (*Figure 2*). The Teflon grooves can be prepared by cutting Teflon tubing in half. A tube slitter as shown in *Figure 3* is recommended for mass production. The tubing should have an inner diameter of close to 1.2 mm to fit the oocytes. The stoppers are made of Teflon rods, 1.2 mm thick, cut into short (about 1.0 mm) pieces; they should fit snuggly into the grooves. It is convenient to attach 3–4 Teflon grooves, each about 2 cm

The Xenopus oocyte cell–cell channel assay

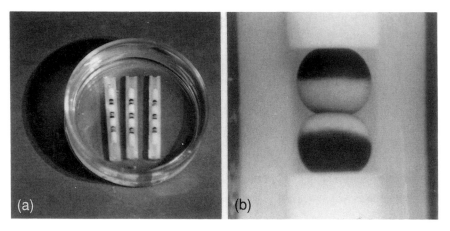

Figure 2. Oocyte pairs mounted for voltage-clamp analysis. (a) Tissue culture dish (35 × 10 mm) with Teflon grooves holding nine oocytes pairs. (b) Oocyte pair held in place by Teflon stoppers. The oocytes are paired with their vegetal poles (the lighter, less pigmented section) in apposition.

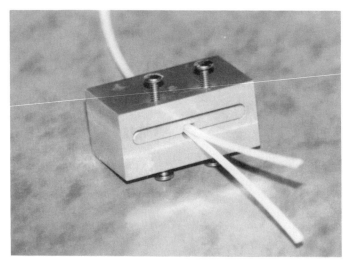

Figure 3. Tubing slicer. The Teflon tubing is pushed into a snugly fitting borehole in a metal block. At the exit the tubing is sliced by a razor blade that is mounted with a shim and setscrews slightly off-centre of the borehole.

long, to 35 × 10 mm tissue culture dishes with sylgard. The sylgard level should be at the rim or slightly below the Teflon grooves. The oocytes are transferred into these dishes filled with OR2 medium and oocytes are gently pushed together to form pairs. The pairs are held in place by the Teflon stoppers. Ten to twelve pairs can easily be mounted in a single dish. Best results are obtained when the vegetal poles of the paired oocytes are facing each other (20).

3. Functional tests for expression of cell–cell channels

Gap junction channels allow the intercellular flux of small ions and molecules with a cut-off at around 1000 daltons. The existence of open channels between paired oocytes can therefore be deduced from either current flow from cell-to-cell or from the transfer of tracer molecules injected into one oocyte to the other.

3.1 Dual voltage-clamp

The method of choice to accurately quantitate the junctional conductance is the dual voltage-clamp (21). A complete discussion of the theory and potential artefacts of the dual voltage-clamp technique are presented in Chapter 8 of this volume ('Patch-clamp analysis of gap junctional currents' by Veenstra and Brink). In this technique, both oocytes of a pair are individually voltage-clamped, i.e. the membrane potential is compared to a set point and via a feedback loop the difference is kept at a minimum. The set point (holding potential) can be chosen to be equivalent to the resting membrane potential (no current flow) or any deviation of it. Each oocyte is impaled with two microelectrodes, one for voltage recording and one for current injection. The electrodes are glass capillaries heated and pulled to give resistances of 1–5 megaohm when filled with 0.5 M K_2SO_4. An Ag/AgCl reference electrode is inserted into a chamber filled with 0.5 M K_2SO_4 which is electrically connected to the bath via an agar bridge. The bath is held at virtual ground.

When both cells are initially held at the same potential, voltage steps applied to one oocyte (ΔV_1) will establish a voltage gradient across the cell junction (V_j). This represents a driving force for a transjunctional current (i_j). To keep the second oocyte of the pair at its holding potential (V_2), the clamp circuit for this oocyte counteracts the disturbance by i_j, by putting out a clamp current that is of the same magnitude but of opposite sign as i_j, therefore, $i_2 = i_j$.

I_1, is the sum of the non-junctional membrane currents i_{m1} and the junctional current i_j. A bath current monitor will see the sum of all currents, i.e. $i_1 + (-i_2) = i_m$, and thus yields information about the non-junctional membrane conductance. Junctional conductance is calculated by $g_j = i_j/V_j$. The junctional conductance, g_j, is determined by: (*i*) number of channels, (*ii*) their open probability, and (*iii*) their unit conductance. It is not possible to determine which of these parameters may underlie a change in macroscopic g_j a priori.

For symmetry test, the voltage steps should be applied to both oocytes of a

The Xenopus oocyte cell–cell channel assay

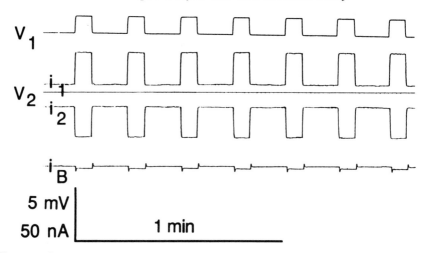

Figure 4. Dual voltage-clamp recording from oocyte pair expressing connexin32. The record was taken 20 h after mRNA-injection and 2 h after pairing. Junctional conductance calculates to 18.7 µS and non-junctional membrane conductance to about 1 µS.

pair successively or alternatingly. A typical recording with the potential of one oocyte being stepped is shown in *Figure 4*.

In practice, the membrane potentials of the two oocytes in a pair may not be identical, and a continuous current may flow through the junction. In that case, the true membrane potentials have to found in an iterative process by minimizing the clamp currents. Potential differences of 5 mV should, however, not be tolerated, in particular if voltage-dependence properties are to be studied. Such pairs should not be scored.

For data acquisition five channels should be available:

- V_1 voltage of oocyte 1
- V_2 voltage of oocyte 2
- i_1 current delivered to oocyte 1
- i_2 current delivered to oocyte 2
- i_B current recorded by the bath current monitor

Voltages and currents may be displayed on an oscilloscope, recorded on a chart recorder or be stored and analysed in digitized form after A to D conversion.

A set-up is shown in *Figure 5* and includes:

- a platform for the recording chamber which can be moved vertically by a rack and pinion mechanism
- four micromanipulators for the voltage and current electrodes
- a device for perfusing the recording chamber (for example, to apply CO_2 or other agents affecting g_j)

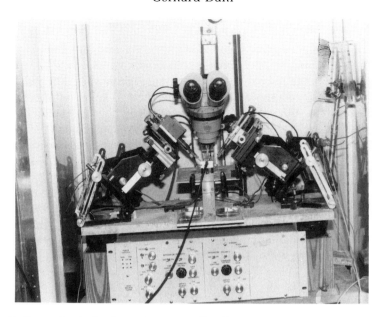

Figure 5. Set-up for dual voltage-clamp recording from oocyte pairs.

- stereo-microscope
- light source (to minimize electrical noise a fibre optic device is recommended)
- voltage-clamp apparatus

The complete set-up is mounted in a cooling cabinet with a temperature range of 5–25°C. This serves two functions: (*i*) the oocytes can be kept continuously at 19°C; (*ii*) the metal enclosing acts as a Faraday cage.

Relatively coarse, inexpensive micromanipulators are adequate to hold the voltage and current electrode. We use four WPI M3301 manipulators which provide coarse X, Y, and Z axis movements as well as a finer Z-axis control for final impalement. The manipulator should be mounted on an adjustable tilting stand (WPI M3 or equivalent) to facilitate electrode changes. The tilting base may be conveniently mounted on a magnetic base for stability.

The voltage-clamp apparatus should be designed for the specific needs for clamping *Xenopus* oocytes. For example, clamp currents in excess of 10 μA must be delivered by the device. Many commercial voltage-clamps are designed for use with single channels, or relatively low membrane currents produced in small cells, and are not appropriate for use with oocytes. Others cannot be used for dual voltage-clamping with a common reference without modification. A version of the apparatus used in the author's laboratory is available from Knight Industries, Miami, Florida.

3.2 Tracer flux

Fluorescent dyes are the most commonly used tracer molecules for monitoring junctional transfer of molecules. Such tracers are difficult to use in *Xenopus* oocytes because of the peculiar optical properties of these cells; for example, heavy pigmentation. In addition, there is some autofluorescence, mainly associated with yolk platelets, which is compounded by the large size of oocytes. As an alternative, radioactive tracer molecules can be employed as long as they fulfil the same criteria as dyes: not being metabolized and not being pumped or leaking through the non-junctional membrane. For example, carrier-free ^{35}S-sulfate can conveniently be used (4). One oocyte of a pair is injected with ^{35}S-sulfate. After an incubation period (30–60 min) the oocytes are carefully separated, using tweezers under a dissecting scope. Oocytes may then be washed and transferred into scintillation vials. Radioactivity is determined with a scintillation counter after the addition of a water-based fluor such as Aquasol (New England Nuclear).

4. Independent verification of expression

One of the major uses of the oocyte cell–cell channel assay will be for structure–function analysis involving mutagenesis. In cases where loss of function is the apparent consequence of a specific mutation, it will be important to demonstrate that the protein is synthesized and undergoes correct post-translational processing. The latter includes correct membrane insertion and intercellular movement. Any of these various steps could be interfered with by the mutation.

4.1 Protein chemistry

Connexins made from synthetic mRNA are so abundant in the oocyte that they can often be visualized by simple metabolic labelling and SDS gel electrophoresis of the oocyte proteins (22). Oocytes are coinjected with mRNA and ^{35}S-methionine, homogenized several hours later, and oocyte proteins analysed by SDS gel electrophoresis. Alternatively, the connexin proteins can be immunoprecipitated before gel electrophoresis (see ref. 22 for details). Radiolabel (10 mCi/ml) may be added directly to the mRNA to be injected. However, best results are usually obtained by increase the label concentration four- to tenfold by lyophilizing in a Speed-Vac (Savant) or equivalent. For gel analysis, the oocyte may be directly homogenized in 100–500 μl of sodium dodecyl sulfate (SDS) sample buffer.

4.2 Immunohistochemistry

Immunohistochemistry may be used to gain information about the cellular location of the newly synthesized connexins. For example, wild-type

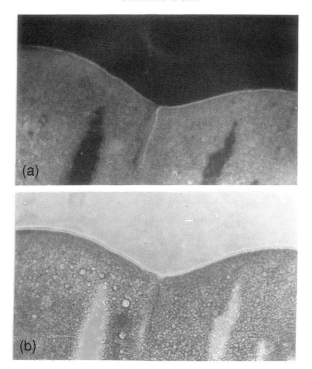

Figure 6. Immunofluorescence (a) and corresponding phase-contrast (b) photomicrographs of frozen sections of oocyte pairs injected with connexin32 mRNA. A rabbit antiserum prepared against a peptide representing the sequence 104–122 of connexin32 was used as primary antibody and FITC-labelled goat-anti rabbit IgG as secondary antibody. In addition to cytoplasmic localization, connexin32 is concentrated in a compartment in or close to the plasma membrane. Both the interface between the two oocytes as well as the non-appositional surface exhibit specific staining.

connexin32 accumulates in a compartment that is within, or extremely close to, the cell membrane (*Figure 6*), presumably forming a pool of precursors from which cell–cell channels can rapidly be formed. If, as a consequence of a specific mutation, only cytoplasm staining without surface labelling should be detected, then an impairment of trafficking could be a probable cause. It is preferable to perform the immunohistochemical analysis on the same oocyte pairs that have been subjected to the measurement of junctional conductance to get a strict correlation. If Teflon troughs (Section 2.6) are used for mounting the pairs, the oocytes can easily be removed from them without tearing the membrane.

Connexin specific antibodies, monoclonal as well as polyclonal, have been generated in various laboratories and have been made available to other researchers. In order to verify that the antibody interaction is specific, standard procedures should be folllowed which include pre-immune serum and

The Xenopus oocyte cell–cell channel assay

competition with the antigenic peptide. In addition, it is convenient to include uninjected oocytes in the same cryosection; for example, hybrid pairs where one oocyte is injected with connexin specific mRNA and the other is not injected.

4.3 Electron microscopy

When mutant connexins are expressed in oocytes it is likely that some mutants will be encountered that make no functional channels but show on the basis of immunohistochemistry normal distribution of the connexin proteins.

In such instances it will be useful to know whether channels are formed that cannot open or whether other steps (for example, docking and formation of the gap junction plaques) are impaired. A discrimination between these possibilities can be done by electron microscopy. Oocyte pairs expressing exogenous connexins exhibit many gap junction structures at the interface that can easily be identified by electron microscopy (EM) (*Figure 7*).

Both for thin-section and freeze-fracture EM techniques, standard procedures for tissue preparation can be employed. However, for freeze-fracturing it is advisable to dissect the interface after fixation of the oocytes with

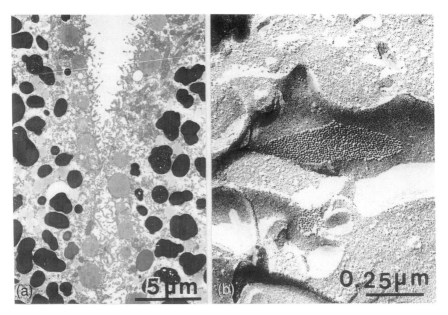

Figure 7. Thin-section (a) and freeze-fracture (b) electron micrographs of interface between paired oocytes injected with connexin32 mRNA. The low magnification thin-section micrograph reveals the elaborate microvillar surface that extends into the contact area. The clustered intramembranous particles typical of gap junctions can be seen in the freeze-fracture electron micrograph (b).

glutaraldehyde. The rationale for this is to increase the chance for the fracture plane to go through the interfacial membranes. It is probably due to the high lipid content of yolk granules and their large size that, in contrast to most other cells, in oocytes the fracture plane prefers a path through the cytoplasm along yolk granules.

5. Discriminating between endogenous and exogenous connexins

Ideally, the host cell for the exogenous protein should not express an endogenous counterpart. Cultured cell lines have been described where cell–cell communication is not detected (5, 6). However, in some cases, the cells are known to express connexins (ref. 26, and D. Paul, personal communication). In these cases, the failure to establish intercellular communication must be due to other factors, such as the inability to assemble or open cell–cell channels. *Xenopus* oocytes, like most cultured cells, express an endogenous connexin (*Xenopus* connexin38; ref. 23) and can display endogenous cell–cell channel activity.

The presence of endogenous connexins or cell–cell channel activity does not necessarily limit the usefulness of a given assay system. The ability to distinguish endogenous from induced cell–cell channels will depend mainly on three factors:

(a) the difference between the levels of induced and endogenous communication;
(b) differences in the characteristic physiological properties of the channels; and
(c) the ability to modulate endogenous communication and/or eliminate the endogenous connexin expression.

In oocytes, the characteristics of the endogenous channels permit their suppression or discrimination from exogenous channels by several simple methods, outlined below.

5.1 To achieve high ratios of exogenous to endogenous coupling

Typically, connexin-specific mRNA induces a junctional conductance three orders of magnitude higher than in uninjected or water-injected control pairs (*Table 1*). Most of the time, endogenous channels are not made in significant numbers. If they exist, they can often be recognized by their characteristic gating properties (*Table 2*). Provided the exogenous connexins have higher affinities to themselves than to the endogenous proteins, the contribution by endogenous proteins to the total junctional conductance and the gating properties of the channels can probably be safely neglected. These conditions

Table 1. Junctional conductance of oocyte pairs in a typical experiment

Time after pairing	Control oocytes	Connexin32 mRNA injected
2 hr	0 (10)	13.16 ± 3.45 (9)
24 hr	0.10 ± 0.02 (10)	120.28 ± 3.27 (9)

Conductances are expressed in microsiemens, and the means (SEM) are given. Connexin32 mRNA was injected 18 h before pairing. After removal of the vitelline layer, oocytes were incubated for 20 min in OR2 containing 10 μg/ml glycine max. Oocytes were washed and then paired in OR2 medium and junctional conductance determined 2 h and 24 h after pairing.

Table 2. Gating properties of various connexin channels expressed in *Xenopus* oocytes

	Endogenous	Connexin38	Connexin32	Connexin43
Sensitivity to				
CO_2	+++	+++	+	+
Ca^{2+} (cytoplasmic)	–	–	(+)	+
Voltage (V_j)	++	++	(+)[a]	–
pH → voltage (V_j)	↓	↓	↑	(–)[b]
Octanol	–	ND	–	–

[a] See *Figure 8*
[b] See ref. 25
[c] Octanol closes cell–cell channels in most other cells by an unknown mechanism. In oocytes the site of action for this compound may not be accessible or may even not exist.

are adequately met for connexin32. Connexin43 is less satisfactory in this regard, as it exhibits significant affinity for endogenous *Xenopus* connexins.

5.2 Time-pairing to avoid endogenous coupling

If pairing is performed with a delay (about 12 h) after injectrion of mRNA, cell–cell channels form rapidly from a pool that accumulated during the incubation period. Endogenous channels do not profit from this delay and form as slowly as without the extra incubation period. Thus, the delayed pairing protocol provides a time window of up to several hours after pairing in which exclusively exogenous channels are present.

5.3 Eliminating endogenous connexins

Expression of proteins can be suppressed by antisense RNA transcription. The antisense strategy has been successfully applied in *Xenopus* oocytes for various proteins, including exogenous connexins (3). More recently, the elimination of the endogenous *Xenopus* connexin38 has been achieved by injection of antsense DNA oligonucleotides (ref. 24, and B. Gimlich, personal communications). After the injection of the antisense DNA oligonucleotides, sufficient time must be allowed for complete turnover of all the endogenous connexins. This appears to require between 24 and 72 h depending on the particular batch of oocytes. Injection of exogenous connexin RNA should be delayed until endogenous connexin is removed.

5.4 Selective inhibition of endogenous coupling

Depending on the purpose of the experiment, the differential sensitivity to cytoplasmic acidification between endogenous and exogenous connexins can be used to selectively block out endogenous channels. For example, perfusing the oocyte pairs with medium saturated with a 10% CO_2–90% air mixture will completely abolish junctional conductance provided by endogenous channels (25). In contrast, channels made from connexin32 or 43 are not

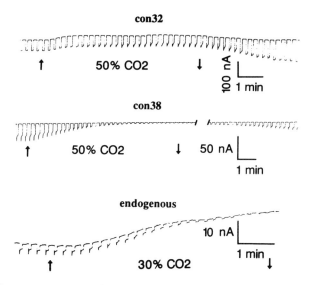

Figure 8. Differential responses of various connexins to cytoplasmic acidification by CO_2. Records were taken 2 hours after pairing for connexin32, and 24 h after pairing for connexin38 and endogenous channels. Only transjunctional currents are shown. CO_2 application is indicated by arrows. Oocytes injected with connexin38 exhibit the same type of response as oocytes expressing endogenous channels only: junctional conductance is abolished by low CO_2 concentrations and the voltage gate that is present at physiological pH becomes almost ineffective. Oocytes expressing connexin32, in response to 50% CO_2, increase junctional conductance and the channels reversibly exhibit voltage-dependent closure. Channels made from connexin43 respond similarly to connexin32 channels except that the appearance of voltage-dependence requires extreme acidification (25).

closed by this procedure. In fact, the macroscopic junctional conductance is slightly increased by an as yet undetermined mechanism.

To close channels formed of connexin32 or connexin43, CO_2 concentrations exceeding 50% are required. Therefore, if the recordings are done in the presence of 10–50% CO_2, only the contribution of the exogenous channels to the junctional conductance will be seen. A potential limitation of this procedure is that cytoplasmic acidification can affect other parameters. For example, acidification renders connexin32 channels much more sensitive to closure by transjunctional voltage gradients (*Figure 8*; ref. 25).

Acknowledgements

Many thanks to Dr R. Werner for sharing a lasting friendship, to E. Levine for his contributions to the assay described in this chapter, to Dr W. Nonner and B. Knight for help in designing and building the voltage-clamp apparatus, and to Katrina Florence for typing the manuscript.

References

1. Beyer, E. C., Paul, D. L., and Goodenough, D. A. (1990). *J. Membr. Biol.*, **116**, 187.
2. Noma, Y., Sideras, P., Naito, T., Bergstedt-Lindquist, S., Azuma, L., Severinson, E., Tanabe, T., Kinashi, T., Matsuda, F., Yaoita, Y., and Honjo, T. (1986). *Nature (Lond.)*, **319**, 640.
3. Dahl, G., Miller, T., Paul, D., Voellmy, R., and Werner, R. (1987). *Science*, **236**, 1290.
4. Werner, R., Miller, T., Azarnia, R., and Dahl, G. (1985). *J. Membr. Biol.*, **87**, 253.
5. Eghbali, B., Kessler, J. A., and Spray, D. C. (1990). *Proc. Natl Acad. Sci. USA*, **87**, 1328.
6. Zhu, D., Caveney, S., Kidder, G. M., and Naus, C. C. G. (1991). *Proc. Natl Acad. Sci. USA*, **88**, 1883.
7. Gurdon, J. B., Lane, C. D., Woodland, H. R., and Marabaix, G. (1971). *Nature (Lond.)*, **233**, 177.
8. Taylor, H. A., Johnson, A. D., and Smith, L. D. (1985). *Proc. Natl Acad. Sci. USA*, **82**, 6586.
9. Dumont, J. N. (1972). *J. Morphol.*, **136**, 153.
10. Siegel, E. (1990). *J. Membr. Biol.*, **117**, 201.
11. Kozak, M. (1987). *Nucleic Acids Res.*, **15**, 8125.
12. Richter, J. D. and Smith, L. D. (1984). *Nature (Lond.)*, **309**, 378.
13. Drummond, D. R., Armstrong, J., and Colman, A. (1985). *Nucleic Acids Res.*, **13**, 7375.
14. Melton, D. A., Krieg, P. A., Rebagliatti, M. R., Maniatis, T., Zinn, K., and Green, M. R. (1984). *Nucleic Acids Res.*, **12**, 7035.
15. Methfessel, L., Witzemann, V., Takaharhi, T., Mishina, M., Numa, S., and Sakmann, B. (1986). *Pflügers Arch.*, **407**, 577.

16. Dahl, G., Levine, E., Rabadan-Diehl, C., and Werner, R. (1991). *Eur. J. Biochem.*, **197**, 141.
17. Dahl, G., Werner, R., and Levine, E. (1988). In *Gap Junctions, Modern Cell Biology* (ed. E. L. Hertzberg and R. G. Johnson), p. 183. Alan R. Liss, New York.
18. Werner, R., Levine, E., Rabadan-Diehl, C., and Dahl, G. (1989). *Proc. Natl Acad. Sci. USA*, **86**, 5380.
19. Levine, E., Werner, R., and Dahl, G. (1991). *Am. J. Physiol.*, **261**, C1025.
20. Levine, E., Werner, R., and Dahl, G. (1988). *Biophys. J.*, **53**, 51a.
21. Spray, D. C., Harris, A. L., and Bennett, M. V. L. (1981). *J. Gen. Physiol.*, **77**, 77.
22. Swenson, K. J., Jordan, J. R., Beyer, E. C., and Paul, D. L. (1989). *Cell*, **57**, 145.
23. Ebihara, L., Beyer, E. C., Swenson, K. J., Paul, D. L., and Goodenough, D. A. (1989). *Science*, **243**, 1194.
24. Barrio, L. C., Suchyna, T., Bargiello, T., Xu, L. X., Roginski, R. S., Bennett, M. V. L., and Nicholson, B. J. (1991). *Proc. Natl Acad. Sci. USA*, **88**, 8410.
25. Werner, R., Levine, E., Rabadan-Diehl, C., and Dahl, G. (1991). *Proc. R. Soc. Lond. B*, **243**, 5.
26. Musil, L. S. and Goodenough, D. A. (1991). *J. Cell Biol.*, **115**, 1357.

8

Patch-clamp analysis of gap junctional currents

R. D. VEENSTRA and P. R. BRINK

1. Introduction: dual whole-cell patch-clamp analysis of gap junctional currents

The difficulty in recording gap junction channel currents and gating activity can largely be attributed to the specialized location of this ion channel. While virtually all other ion channels are single membrane structures which permit the passage of ions between the extracellular space and the cytoplasm or between the cytoplasm and an internal organellar compartment, gap junctions link the cytoplasms of two adjacent cells. Hence, the direct measurement of junctional currents requires the ability to independently control the membrane voltage of each member of a coupled pair of cells or direct voltage-clamp recording from gap junction membranes. The two-cell voltage-clamp approach was first accomplished using two microelectrodes impaled into each cell, one for passing current and one for recording voltage (1). The two-microelectrode voltage-clamp technique is well suited for use with cells of large diameter (above 20 μm), while the patch-clamp technique offers the advantage of requiring only a single electrode for intracellular current injection and voltage recording from small diameter cells (below 20 μm) (2). The use of the double whole-cell recording (DWCR) procedure for quantitative analysis of macroscopic and single-channel gap junctional currents is now widespread and has yielded valuable new information about channel conductances and the regulation of electrical coupling in a variety of tissues (3).

It is the purpose of this chapter to describe the application of the DWCR technique to the analysis of junctional currents from pairs of electrically coupled cells. This will be accomplished by deriving the equations necessary for calculation of junctional conductance and demonstrating the influence of varying series, input, and junctional resistances on these measurements, using a model two-cell circuit designed to have properties similar to suitable biological preparations. In this manner, it is hoped that the advantages and drawbacks of the DWCR technique to the analytical analysis of gap junction currents will be presented in a clear and orderly fashion. Later in this chapter, the direct patch analysis of gap junction membranes will be presented and compared to the DWCR technique.

2. Preparation of cell pairs

As for any patch clamp experiment, the desired starting point is an ample supply of viable cells with a 'clean' plasma membrane. Quite often, primary cultures of cells are obtained by enzymatic dissociation of tissues such as liver or heart, although established cell lines may also be used. Since there is a wide variety of cell types and tissues available for electrophysiological examination, a description of cell culture procedures is not possible. Instead, the investigator should review specific references for cell isolation pertaining to the tissue of interest, although some general information about the preparation of cells for patch-clamp studies is available (4, 5). The basic method for dissociation of cells from a tissue involves excision of the tissue and removal of blood, dispersion of cells by enzymatic digestion combined with mechanical agitation, and resuspension of cells in physiological saline or culture medium.

The most straightforward approach to obtaining cell pairs for electrophysiological examination of junctional communication is to utilize cell pairs which have remained intact throughout the dissociation procedure. Alternatively, cells may also be manipulated into contact and incubated overnight (6) or recorded from immediately after contact to observe the development of electrical coupling (7). Newly formed cell pairs typically have lower junctional resistances than pairs which have remained intact during the dissociation procedures, although junctional conductances can vary from 0 to greater than 30 nanosiemens (nS) in intact cell pairs. For reasons that will become more apparent later on in this chapter, high cell input and junctional resistances are advantageous for quantitative analysis of junctional currents. Most small diameter cells have input resistances of 1 to 10 gigaohms (GΩ), owing to their small membrane surface area. For example, a 20 µm diameter spherical cell with a specific membrane resistance of 12 kΩ-cm^2 will have an input resistance of approximately 1 GΩ. Resting cardiac adult mammalian ventricular myocytes have a specific membrane resistance of approximately 10 kΩ-cm^2. For most cell types, the specific membrane resistances should be even higher.

2.1 Achieving the dual whole-cell recording (DWCR) configuration

To perform DWCR, you will need the following equipment:

- two patch-clamp amplifiers
- a voltage stimulator
- two three-axis micromanipulators with fine (< 1 µm) and coarse movement controls

- an inverted phase-contrast light microscope with 40× objective
- a vibration isolation table
- a storage oscilloscope (digital)
- a tape-recorder (digital)

In order to obtain high resistance GΩ seals, the cells must be bathed in a protein-free saline solution. Next, patch pipettes with resistances of 2–5 MΩ are introduced into the bath under positive pressure and manipulated into close contact with both cells. The patch pipettes should be filled with a suitable high K^+, low Na^+, Ca^{2+}/EGTA, KOH/Hepes buffered solution to mimic the intracellular environment of mammalian cells (8). Quite often, Cl^- is replaced by other less mobile anions such as glutamate or aspartate. ATP is also frequently added to whole-cell pipette solutions to prevent metabolic rundown of the cells, which can have a dramatic impact on non-junctional membrane resistances. It should be noted that addition of 3 mM Na_2ATP lowers the pH of a 10 mM Hepes buffered solution (pH 7.2) by 0.4 to 0.6 pH units, so the pH should be titrated to the desired level with 1 N KOH or one can increase the Hepes concentration to provide a greater buffering capacity to the pipette solution. Upon making initial contact with the cell surface, positive pressure is released from the electrodes and GΩ seal formation is achieved with the transient application of slight negative pressure.

At this point, the seal resistance (R_s) is measured by applying a 10–20 mV short-duration (e.g. 10 msec) pulse to each pipette. R_s is determined by dividing the voltage pulse amplitude by the pipette current (for example, R_s = 20 mV/20 pA = 1 GΩ). The accuracy of this measurement is based on the assumptions that the membrane patch resistance is much higher than the seal resistance and that the patch currents are negligible and do not alter the intracellular potential (9). A second critical procedure to implement at this time is to compensate for the electrode capacitance before disrupting the membrane patch. Electrode capacitance can be minimized by coating the electrode tip with Sylgard® resin (10). Coating the portion of the electrode that comes into contact with the bath solution reduces the electrode capacitance to 1–2 pF, a value which can be easily compensated for by the fast transient compensation circuit of most patch-clamp amplifiers. This adjustment is important if series resistance compensation is to be used to reduce errors in clamping the cell membrane potential (V_m) (11). Elimination of the fast capacitive transient also serves as an aid for monitoring disruption of the membrane patch to achieve the whole-cell configuration and determining the magnitude of series resistance from the whole-cell capacitive transient (see Section 3.1.1 below). One useful procedure for monitoring the patch break is to apply a 1 mV, 10 msec voltage pulse to each electrode while applying negative pressure to the electrode. Prior to patch break, the current signal should be a flat line. Disruption of the membrane patch is readily observed by the reappearance of a fast capacitive transient, this time due to the cell input capacitance (C_{in}). Hopefully, the holding current does not increase dramatically with patch break, indicating that the integrity of the GΩ seal was maintained.

3. Equivalent circuit for the dual whole-cell configuration

Once GΩ seals and patch break have been achieved on a cell pair, the two-cell preparation is represented by the equivalent resistive circuit shown in *Figure 1*. Each whole cell recording configuration is represented by an electrode resistance (R_{el}) in series with a parallel combination of seal and membrane resistances (R_s and R_m). The junctional resistance (R_j) links the interior of both cells. There are three important features of the equivalent resistive circuit which must be considered when performing R_j measurements. First, the parallel combination of R_s and R_m defines the input resistance (R_{in}) of the preparation according to the following equation:

$$R_{in} = [(R_s \times R_m)/(R_s + R_m)]. \qquad [1]$$

This R_{in} value provides the non-junctional resistance that lies between the cell interior and external (bath) ground and is important in determining

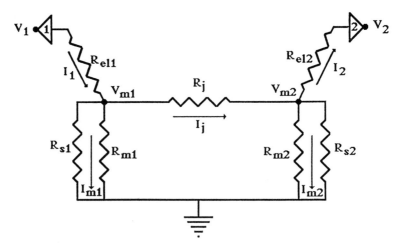

Figure 1. Equivalent resistive circuit for the double whole-cell recording (DWCR) configuration. R_{el1} and R_{el2} are the resistance of the two patch electrodes; R_{m1} and R_{m2} are the membrane resistances of cells 1 and 2; R_{s1} and R_{s2} are the seal resistances of the electrodes to their respective cell membranes; R_j is the junctional resistance; V_1 and V_2 are the command potentials supplied by patch-clamp amplifiers 1 and 2; V_{m1} and V_{m2} are the membrane voltages of cells 1 and 2; I_{m1} and I_{m2} are the membrane currents flowing across the input resistance of each cell; I_1 and I_2 are the holding currents for each cell supplied by their respective amplifiers; and I_j is the junctional current. The current arrows are drawn to indicate the direction of current flow following a voltage step applied to cell 1 from the initial conditions where $V_1 = V_2$.

membrane voltage (V_m). According to Ohm's law, V_m equals the product of the membrane current (I_m) and input resistance, or $V_m = R_{in} \times I_m$. Second, R_{el} is in series with both R_{in} and R_j of the two cell preparation and may contribute significantly to both junctional and non-junctional resistance measurements, depending on the relative value of R_{el} compared to R_{in} and R_j. Third, R_j is isolated from external ground by R_{in} of both cells. This means that any change in junctional current (I_j) does not lead to a change in total holding current ($I_1 + I_2$) for the two-cell preparation, provided that V_{m1}, V_{m2}, R_{in1}, and R_{in2} have remained constant during the recording period.

When both patch-clamp amplifiers are in voltage-clamp mode, command potentials (V_1 and V_2) are applied to each cell in an effort to control V_m of each cell. From the circuit diagram, it is apparent that V_m will deviate from V by an amount equal to the product of R_{el} and the holding current (I), or

$$V_m = V - (R_{el} \times I). \qquad [2]$$

Hence, $V_m = V$ occurs only when R_{el} is negligible. This condition is optimized by using low-resistance patch electrodes ($R_{el} < 5$ MΩ) and high input resistance cells ($R_{in} > 2$ GΩ). The actual transjunctional voltage (V_j) is therefore equal to the difference between V_{m1} and V_{m2}. When using patch electrodes, it is important to remember that one cannot directly measure V_m unless separate voltage microelectrodes are used, as in the double microelectrode voltage-clamp technique (1).

3.1 Derivation of equations for junctional conductance measurements

In the simplest of terms, R_j can be calculated according to Ohm's law

$$R_j = V_j/I_j = 1/g_j, \qquad [3]$$

where g_j is the junctional conductance. The usual approach to measuring R_j is to begin with a set of initial conditions where $V_1 = V_2$ ($V_j = 0$ mV) and the holding currents I_1 and I_2 are obtained under conditions where there is no net junctional current flow ($I_j = 0 = V_j/R_j$). Hence, $I = I_m$ for each cell. To measure R_j, a voltage pulse (ΔV) is applied to only one of the cells, thus creating a non-zero V_j and a net I_j to flow between the cells. The current arrows in *Figure 1* are drawn to indicate the direction of current flow following a voltage step (ΔV_1) applied to cell 1. Under these conditions, $V'_1 = V_1 + \Delta V_1$, $V'_2 = V_2$, $I'_1 = I_1 \times \Delta I_1$, and $I'_2 = I_2 + \Delta I_2$. V'_1, V'_2, I'_1, and I'_2 are all directly measured parameters. Next, we need some sort of convention for defining V_j and I_j. Most prefer to subtract V_m of the non-pulsed cell from the V_m of the cell to which ΔV was applied (6, 7, 12). The opposite convention has been used elsewhere and is of little consequence to the final results, for reasons that will become evident later in this chapter (13, 14, 15, 16). From equation 2 we obtain the following expression:

$$V_j = V'_{m1} - V'_{m2} = V'_1 - V'_2 - [(I'_1 \times R_{el1}) - (I'_2 \times R_{el2})]$$
$$= \Delta V_1 - [(I'_1 \times R_{el1}) - (I'_2 \times R_{el2})]. \quad [4]$$

Both ΔI_1 and ΔI_2 have junctional (I_j) and membrane (I_m) current components. Given that V_{m1} has been intentionally altered by an amount equal to $\Delta V_1 - (\Delta I_1 \times R_{el1})$, ΔI_{m1} will be significant ($= [\Delta V_1 - (\Delta I_1 \times R_{el1})]/R_{in1}$). If one considers all of the possible pathways for a current injected into cell 1 to flow, then $\Delta I_1 = \Delta I_{m1} + I_j$. It is best to obtain a measure of I_j from cell 2 since V'_2 has remained constant and any ΔI_{m2} should be negligible. In Section 3, it was stated that since R_j is isolated from ground by R_{in} of each cell, any change in I_j must not lead to a change in the sum of I_1 and I_2 (under conditions of constant V_m and R_{in}). Assuming V_j remains constant for a given ΔV_1 then $\Delta I_{m1} = \Delta I_{m2} = 0$ and $\Delta I_1 - I_j = \Delta I_1 + \Delta I_2 = 0$. Hence, $\Delta I_1 = I_j = -\Delta I_2$. The concept that I_j must be recorded as currents of equal amplitude and opposite polarity by the two voltage-clamp circuits has been referred to as the 'equal and opposite' criterion, and serves as a useful criterion for distinguishing junctional current from non-junctional membrane currents (1, 14). In practice, the actual value of $\Delta I_{m2} = I_j \times [R_{el}/(R_{el} + R_{in2})]$. From the above two expressions, it follows that:

$$I_j = -\Delta I_2 \times (1 + R_{el2}/R_{in2}). \quad [5]$$

Combining equations 3, 4, and 5, one obtains the following expression:

$$R_j = 1/g_j = \{\Delta V_1 - [(I'_1 \times R_{el1}) - (I'_2 \times R_{el2})]\}/ \quad [6]$$
$$[-\Delta I_2 \times (1 + R_{el2}/R_{in2})].$$

Values for ΔV_1, I'_1, I'_2, and ΔI_2 are easily obtained from direct measurements of the current and voltage outputs of the patch-clamp amplifiers and R_{in} and R_{el} should be determined during the course of each experiment. If ΔV was applied to V_2 instead of V_1, a similar set of equations would be derived.

3.1.1 Estimation of electrode series resistance

It should be apparent from equations 2 and 5 that accurate determination of I_j and V_j requires R_{el} to be small relative to R_{in}. R_{el} should be routinely measured prior to GΩ seal formation by measuring the pipette current in response to a 100–200 μV voltage pulse applied to each electrode. As stated above, electrode resistances of 2–5 MΩ are standard. Patch-clamp amplifiers usually contain a series resistance compensation circuit which corrects for the voltage drop across R_{el} by applying a variable voltage to the command voltage, V (11). Because the series resistance compensation circuit depends upon the accuracy of the electrode capacitance cancellation circuit and can cause positive feedback oscillations or distortion of the clamped membrane potential unless properly adjusted, this circuit is not usually employed when using cells with $R_{in} > 1$ GΩ.

R_{el} is likely to increase following disruption of the membrane patch due to partial occlusion of the electrode tip. This necessitates being able to estimate the series resistance in the whole-cell configuration. R_{el} can be estimated using the series resistance compensation circuit, although this value will be underestimated if only partial compensation can be achieved. Provided that the fast capacitive transient of the electrode was properly cancelled upon GΩ seal formation as mentioned above, one can obtain an estimate of R_{el} from the whole-cell capacitive transient. When a ΔV is applied to a cell, the membrane capacitance (C_m) is charged through the series resistance. Hence, the capacitive current transient (I_{cap}) will have an amplitude of $\Delta V/R_{el}$ and an exponential decay time-constant approximately equal to $R_{el} \times C_m$ (11). Measuring the amplitude of I_{cap} during the simultaneous application of a ΔV (= 10 mV) to a pair of chick ventricular myocytes has been used as a convenient method for determining R_{el} in the DWCR configuration without employing series resistance compensation (14).

An example of this approach to the estimation of series resistance in the DWCR configuration is shown in *Figure 2*. A model circuit similar to that shown in *Figure 1* was assembled from resistors and capacitors having the following values: R_{el1} = 20 MΩ, R_{el2} = 10 MΩ, R_{s2} = 10 GΩ, R_{m2} = 1 GΩ, and R_j = 1 GΩ. R_{s1} and R_{m1} were substituted for by a single 1GΩ resistor to represent R_{in1} and a 5 pF capacitor was added in parallel to R_{in1} and R_{in2} to represent C_{m1} and C_{m2} of the preparation. In *Figure 2A* and *2B*, R_{el1} and R_{el2} were connected directly to ground and a 20 mV pulse was simultaneously applied to both resistors. Initially, a fast capacitive current could be recorded by each amplifier which settled to a steady-state level in <200 μsec. The steady-state I_1 and I_2 currents measured 928 and 1758 pA and correspond to a calculated R_{el1} = 21.6 and R_{el2} = 11.4 MΩ. Both capacitive signals could be cancelled using the fast capacitive compensation circuit of each patch-clamp amplifier (*Figure 2A, B*; compensated). When attached to the circuit described above, whole-cell capacitive currents (*Figure 2c*) were obtained during the simultaneously applied 20 mV pulse. Calculation of R_{el} from the peak of each transient produced values of R_{el1} = 20 mV/950 pA = 21.1 MΩ and R_{el2} = 20 mV/1096 pA = 17.9 MΩ. Integrating the area under each capacitive transient, the input capacitances (C_{in}) were measured at 9.0 and 7.8 pF, respectively. Decay time-constants were obtained from a single exponential fit of each transient and R_{el} estimates using the product of C_{in} and R_{el} produced estimates of R_{el1} = 9 pF × 169 μsec = 18.6 MΩ and R_{el2} = 7.8 pF × 99 μsec = 12.7 MΩ. It appears from this demonstration that integrating C_{in} and obtaining the decay constant for each whole-cell capacitive transient may be a more reliable measure of R_{el}, since the amplitude of I_{cap} is most sensitive to the amount of fast capacitance compensation applied.

3.1.2 Estimation of cellular input resistance

From equation 6, it is apparent that the R_{in} of the cell being held at a constant

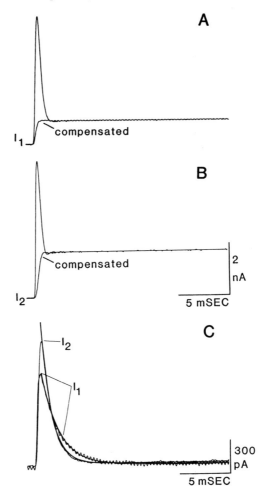

Figure 2. Estimation of electrode resistance from whole-cell capacitive transients. (**A**) Capacitive current transient elicited by a 20 mV voltage pulse before and after capacitance neutralization of a 20 MΩ resistor, R_{el1}, connected to ground. Steady-state current was 928 pA. (**B**) Same as in panel A for a 10 MΩ resistor representing R_{el2}. Steady-state current was 1758 pA. (**C**) Whole cell currents obtained from two coupled circuits with $R_{in1} = 1$ GΩ, $R_{in2} = 0.91$ GΩ ($R_{m2} = 1$ GΩ, $R_{s2} = 10$ GΩ), $R_j = 1$ GΩ, and cell membrane capacitances ($C_{m1} = C_{m2}$) = 5 pF. The solid lines are single exponential fits of the capacitive current decay with time-constants of −169 and −99 μsec. All currents were digitally sampled at 500 kHz after low-pass filtering at 10 kHz.

potential directly affects the I_j measurements. Somewhat less apparent is the fact that R_{in} of both cells is important, since the lower R_{in} becomes, the greater I'_1 and I'_2 become and the greater the series resistance errors become in determining V_j. From equation 1, we know that R_{in} is equal to the parallel combination of R_s and R_m and, although we cannot independently measure these two values in the whole-cell configuration, one can directly measure R_{in} for each cell of a pair. This can be achieved by simultaneously stepping V_1 and V_2 to a series of identical potentials (i.e. -100 to $+60$ mV) and measuring the holding currents (I_1 and I_2) at each voltage. When calculating R_{in}, it is important to note that V should be equal to the difference between the applied command potential and the resting potential (V_r) for the cell, or $R_{in} = (V - V_r)/I$. The resting potential is the voltage at which the holding current is zero, which is not likely to occur at 0 mV unless identical intracellular and extracellular solutions are used or unless R_s has become leaky (< 1 GΩ). As an alternative to calculating R_{in} at a single voltage, one may plot I as a function of V and report R_{in} as the inverse of the chord conductance (17) over the linear portion of the $I-V$ relationship. This approach is useful since it permits the investigator to choose a range of potentials to use during the course of an experiment where R_{in} should be high and I_m will vary linearly with V_m.

3.1.3 Estimation of junctional resistance or conductance

If the whole-cell currents and series resistances can be kept small, then equation 6 reduces to the more familiar expression of:

$$R_j = (V_j/I_j) = -(\Delta V_1/\Delta I_2) \quad [7]$$

when ΔV is applied to cell 1. As mentioned above, V_j is usually defined as $V'_1 - V'_2$ and $I_j = -\Delta I_2$ (6, 7, 12). Regardless of which convention is used, the current or voltage being defined as negative, it should be obvious that the final expression is the same (equation 7). The only difference between the two approaches lies in the transposition of the voltage axis, which would become significant if directional differences in R_j were present (i.e. unidirectional rectification). Equation 7 is the expression which is most commonly used to estimate R_j and I will often refer to R_j values obtained from this equation as the conventional R_j estimate. Hopefully, one can now better understand the critical assumptions which underly the use of equation 7 in calculating R_j (or g_j) using the DWCR approach. High whole-cell currents and high series resistances will lead to significant underestimations of R_j and for these reasons, the DWCR technique is best suited to cell pairs with high input and junctional resistances. These limitations of the DWCR approach will now be demonstrated using a model circuit configured as shown in *Figure 1*. For the circuit demonstrations to follow, these additional items were required:

- one digital voltmeter (0.1 mV sensitivity)
- one breadboard
- two 5 pF capacitors
- six 10 MΩ (5%) resistors, two pair wired in series
- two 100 MΩ (1%) resistors

- two 500 MΩ (5%) resistors
- three 1 GΩ (2%) resistors
- one 10 GΩ (5%) resistor

4. Analysis of junctional currents

As a practical approach to the analysis of junctional currents using the DWCR technique, the validity of the assumptions used to derive equation 7 will be examined using a model DWCR circuit with resistive and capacitive values similar to suitable biological preparations. For simplicity, R_{s1} and R_{m1} of *Figure 1* have been combined into a single 1 GΩ resistor representing R_{in1} and membrane capacitances for each cell (C_{m1} and C_{m2}) are represented by a 5 pF capacitor placed in parallel with R_{m1} and R_{m2}. To begin, lets assume we have a relatively 'ideal' preparation and go through the procedure of calculating R_j using equations 6 and 7 to determine the accuracy of the DWCR method under optimal conditions.

4.1 Ideal preparation

The initial set of conditions for the model circuit are defined in *Table 1*. This circuit is considered to be an ideal preparation for analysis of R_j by the DWCR method since R_{el} of both cells is only 1% of R_{in} and R_j, and the R_{in} and R_j values are approximately 1 GΩ or more. In this example, there is a slight asymmetry in the R_{in} values since the 10 GΩ R_{s2} in parallel with R_{m2} causes a modest reduction in R_{in2}.

4.1.1 Input resistance measurements

R_{in1} and R_{in2} can be directly measured by simultaneously stepping V_1 and V_2 from the resting potential for the DWCR circuit (= 0 mV since no battery is present) to a new potential as described above (Section 3.1.2). *Figure 3A* shows the whole-cell currents (I_1 and I_2) obtained during a −40 mV ($\Delta V_1 = \Delta V_2 = -40$ mV) pulse applied simultaneously to cells 1 and 2. Whole-cell capacitive transients were present in both current traces during the onset and termination of the ΔV pulse and the steady-state holding currents, I'_1 and I'_2, measured −38.9 and −39.7 pA, respectively. Since ΔV is the same for V_1 and

Table 1. Circuit values for an ideal preparation

	R_{el} (MΩ)	R_s (GΩ)	R_m (GΩ)	R_{in} (GΩ)	C_m (pF)	R_j (GΩ)
Cell 1	10	–	–	1	5	1
Cell 2	10	10	1	0.91[a]	5	1

[a] Calculated using equation 1.

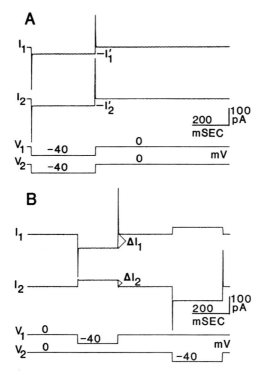

Figure 3. Determination of input and junctional resistances for ideal DWCR circuit. (**A**) Whole-cell currents, I_1 and I_2, measured 0 pA prior to the onset of a -40 mV pulse applied simultaneously to V_1 and V_2 from a holding potential of 0 mV. I'_1 and I'_2 measured -38.9 and -39.7 pA and R_{in1} and R_{in2} estimates of 1.03 and 1.01 GΩ were obtained from these records. Note the presence of capacitive transients in both current traces. (**B**) Alternate application of a -40 mV pulse to V_1 and V_2 produces ΔI_1 and ΔI_2 records of different magnitudes and opposite polarity. Note that only the current record of the pulsed cell has a capacitive transient indicative of a change in membrane potential. The ΔI signal of the non-pulsed cell is approximately equal to I_j, while ΔI of the pulsed cell is equal to $\Delta I_m - I_j$.

V_2, ΔI_1 and ΔI_2 have the same polarity. Based on the equation $R_{in} = \Delta V/\Delta I$, R_{in1} and R_{in2} were calculated to be 1.03 and 1.01 GΩ. Taking into account the expected voltage drop across R_{el1} and R_{el2} of 0.39 and 0.40 mV (see equation 2) alters the calculated R_{in1} and R_{in2} by only -1% in each case. Hence, under these conditions, it is safe to assume that $V_m = V$.

4.1.2 Junctional resistance measurements

If ΔV is applied alternately to V_1 and V_2, current records like those illustrated in *Figure 3B* are obtained. Initial observations reveal two distinct differences in the whole-cell current responses when V_1 and V_2 are stepped alternately instead of simultaneously. First, the capacitive current transients are observed only in the whole-cell current record of the pulsed cell and, second, I_1 and I_2 are of opposite sign. Capacitive currents should be elicited only when V_m is changing and the absence of a capacitive current serves as a good indicator of proper voltage control of the nonpulsed cell during DWCR experiments.

When V_1 is stepped to -40 mV with V_2 constant at 0 mV, $-\Delta I_2$ is assumed to be the direct measure of I_j while ΔI_1 should equal $\Delta I_{m1} + I_j$. For the example shown in *Figure 3B*, $\Delta I_2 = 38.5$ pA and $\Delta I_1 = -78.8$ pA. From these measurements, it follows that $\Delta I_{m1} = -40.3$ pA, which is comparable to the value of -38.9 pA obtained during the R_{in1} measurement. When V_2 was stepped to -40 mV with V_1 constant at 0 mV, the exact opposite results were obtained, as shown in *Table 2*. Therefore, in either case R_j is estimated to be -40 mV/-38.5 pA or 1.04 GΩ. This conventional R_j estimate is quite acceptable given that the specified value was 1 G$\Omega \pm 2\%$ and our own direct measurements of the actual resistor value were only accurate to within 4% of 1 GΩ. The actual values for V_1, V_{m1}, V_2, V_{m2}, and V_j were also directly measured using a digital voltmeter and matched the predicted (equations 2 and 4) electrode voltage drops for the pulsed and non-pulsed cells, resulting in a V_j of -38.8 mV. So even under conditions where R_{el} is equal to 1% of R_{in} and R_j, the actual V_j is 3% less than the applied ΔV due to a 2% and 1% drop in V_m of the pulsed and non-pulsed cells. Taking into account the actual V_j and I_j of -38.9 pA (equation 5), the corrected R_j estimate (equation 6) was 0.996 GΩ. Assuming the actual resistor value is exactly 1 GΩ, the error in the R_j estimate obtained from equation 6 was reduced tenfold compared to the R_j estimate obtained with equation 7.

Table 2. Analysis of the ideal circuit under dual voltage-clamp conditions.

	ΔV (mV)	$\Delta V_m{}^a$ (mV)	$V_j{}^b$ (mV)	ΔI (pA)	$I_j{}^c$ (pA)	$R_j{}^{d,e}$ (MΩ)	$R_j{}^{e,f}$ (MΩ)
Cell 1	-40	-39.2 $\{-39.2\}$	-38.8 $\{-38.8\}$	-78.8	—	—	—
Cell 2	0	-0.3 $\{-0.4\}$	—	38.5	-38.9	1039 $\{+3.9\%\}$	998 $\{-0.2\%\}$
Cell 1	0	-0.3 $\{-0.4\}$	—	38.5	-38.9	1039 $\{+3.9\%\}$	999 $\{-0.1\%\}$
Cell 2	-40	-39.2 $\{-39.2\}$	-38.8 $\{-38.8\}$	-77.0	—	—	—

[a] Values in brackets calculated using equation 2.
[b] Values in brackets calculated using equation 4.
[c] Values in brackets calculated using equation 5.
[d] R_j calculation made using equation 7.
[e] Values in brackets are % error from specified R_j of 1000 MΩ.
[f] R_j calculation made using equation 6.

4.2 Effects of seal resistance on junctional resistance measurements

The initial conditions set forth above assumed R_s to be 10 GΩ, typical of most experimental conditions. On occasion, the GΩ seal of one cell may suddenly decrease in value during the course of an experiment and further alter the accuracy of the R_j measurements. To illustrate this event, R_{s2} of the model circuit was reduced to 500 MΩ while all other resistor values remained the same as listed in *Table 1*. The net effect of this change is a reduction in R_{in2} to 333 MΩ, or approximately one-third of R_{in1}. Now we have created a set of circumstances where the effect of alternately pulsing cells 1 and 2 may have different effects on the R_j measurements. If V_1 is stepped to -40 mV with V_2 held constant at 0 mV as before, the same V_{m1}, V_{m2}, and V_j values are obtained since R_{in1} and R_j were unchanged as shown in *Table 3*. This means that V_j will be 3% less than ΔV_1 for the above-mentioned reasons. These values were again confirmed by direct voltmeter measurements. However, ΔI_2 measured only 36.8 pA under these circumstances, or 1.7 pA less than before. This produced a conventional R_j estimate (equation 7) of 1087 MΩ, which was 8.7% higher than the specified R_j value. According to equation 5, I_j should have been 37.9 pA and the corrected R_j estimate (equation 6) was 1025 MΩ, or only a 2.5% over-estimation of R_j. So an additional reason for the overestimation of R_j when using equation 7 in this case is the 3% underestimation of I_j as $-\Delta I_2$.

As a worst case scenario, we further reduced R_{s2} to 100 MΩ, thus producing an R_{in2} of 91 MΩ. Again applying a ΔV_1 of -40 mV, V_{m1}, V_{m2}, and V_j were essentially unchanged because R_{in1} and R_j were kept constant. Under these conditions, ΔI_2 measured 35.0 pA and the conventional R_j estimate was 1143 MΩ (14.3% overestimation of R_j). Equation 5 produced an I_j of -38.9 pA and the corrected I_j estimate was 1001 MΩ. According to these calculations, assuming that $-\Delta I_2 = I_j$ produced an 11% underestimation of I_j which must be accounted for by an increase in ΔI_{m2} ($= I_j - \Delta I_2$) during the ΔV_1 pulse. When ΔV is applied to cell 1, equation 5 can be rearranged as follows:

$$I_j = -\Delta I_2 \times \{1 + [(R_{el2}/R_{m2}) + (R_{el2}/R_{s2})]\}. \qquad [8]$$

This expression provides a better understanding for how changes in R_s influence the I_j measurements. There are two important considerations to be made here. First, the actual determining factor for how much R_s will alter I_j measurements also depends on R_{el}. Second, it should be apparent from this expression that R_m will affect R_j measurements in an identical manner. In practice, one cannot measure R_s and R_m separately in the WCR configuration, so R_{in} must be used instead as previously mentioned.

What happens if $\Delta V = -40$ mV is applied to V_2 instead of V_1 as illustrated above. Applying equation 5 to cell 1, one would predict that $I_j = -\Delta I_1 + (0.01 \times -\Delta I_1)$ since R_{in1} is 100 times higher than R_{el1}. So the I_j measurements

Patch-clamp analysis of gap junctional currents

Table 3. Analysis of seal resistance changes on DWCR junctional resistance measurements.

	ΔV (mV)	ΔV_m[a] (mV)	V_j[b] (mV)	ΔI (pA)	I_j[c] (pA)	R_j[d,e] (MΩ)	R_j[e,f] (MΩ)
$R_{s2} = 500$ MΩ							
Cell 1	−40	−39.2 {−39.2}	−38.8 {−38.8}	−78.8	—	—	—
Cell 2	0	−0.4 {−0.4}	—	36.8	−37.9	1087 {+8.7%}	1025 {+2.5%}
Cell 1	0	−0.3 {−0.4}	—	36.8	−37.2	1087 {8.7%}	1027 {−0.1%}
Cell 2	−40	−38.5 {−38.5}	−38.1 {−38.1}	−147.0	—	—	—
$R_{s2} = 100$ MΩ							
Cell 1	−40	−39.2 {−39.2}	−38.9 {−38.9}	−77.0	—	—	—
Cell 2	0	−0.3 {−0.4}	—	35.0	−38.9	1143 {+14%}	1001 {+0.1%}
Cell 1	0	−0.3 {−0.4}	—	35.0	−35.4	1143 {+14%}	1007 {+0.7%}
Cell 2	−40	−35.7 {−35.9}	−35.6 {−35.6}	−406.0	—	—	—

[a] Values in brackets calculated using equation 2.
[b] Values in brackets calculated using equation 4.
[c] Values in brackets calculated using equation 5.
[d] R_j calculation made using equation 7.
[e] Values in brackets are % error from specified R_j of 1000 MΩ.
[f] R_j calculation made using equation 6.

should be accurate to within 1% of the actual value under these conditions. The ΔI_1 values were 36.8 and 35.0 pA for $R_{s2} = 500$ and 100 MΩ, respectively and the calculated I_j values were −37.2 and −35.4 pA. Equation 7 provided an R_j estimate of 1087 and 1143 MΩ for $R_{s2} = 500$ and 100 MΩ, which are identical to the values obtained when V_1 was stepped. Assuming the I_j measurements are accurate under these circumstances, the cause for the symmetrical R_j estimates when alternately stepping V_1 and V_2 with a low R_{s2} must be due to alterations in V_j. Because R_{in2} was reduced to 333 or 91 MΩ by the reductions in R_{s2} to 500 or 100 MΩ, ΔI_2 increased to −147 or −406 pA. From equation 2, one would predict the voltage drop across R_{e12} to be −1.5 or −4.1 mV, respectively. Thus, $V_{m2} = -38.5$ mV when $R_{s2} = 500$ MΩ and

-35.9 mV when $R_{s2} = 100$ MΩ. Because ΔI_1 varied by only -1.8 pA between the two cases, the resulting changes in V'_{m1} were less than 0.05 mV. According to equation 2, $V_{m1} = -0.4$ mV for both cases and the calculated V_j was 38.2 or 35.6 mV for $R_{s2} = 500$ or 100 MΩ. All of the estimated voltages were confirmed by direct measurement of V_1, V_{m1}, V_2, V_{m2}, and V_j with 0.1 mV resolution. Given these actual V_j values, the corrected R_j estimates were 1026 and 1007 MΩ for $R_{s2} = 500$ and 100 MΩ, respectively. Thus, the errors in applying equation 7 to the estimation of R_j when stepping the 'leaky cell' are due to the additional drop in V_m caused by the increase in holding current which must flow across R_{el}.

The final conclusion to be drawn from this demonstration is that reductions in R_{in} caused by a 'leaky GΩ seal' produces errors in the conventional R_j estimates obtained with equation 7. The magnitude of the error is identical whether a voltage pulse is applied to the 'leaky' or non-leaky cell, resulting in symmetrical R_j measurements (when $R_j \geq R_{in}$). The culprits are the underestimation of I_j obtained in the leaky cell when the partner cell is stepped and the over-estimation of V_j when the leaky cell is stepped. The respective errors in the partner cell are negligible, provided that R_{el} is small relative to R_{in} and R_j. These systematic errors can be avoided by using equation 6 to estimate R_j.

4.3 Effects of membrane resistance on junctional resistance measurements

There is no need for an additional demonstration or discussion of the effects of R_m on R_j measurements, since the answer has already been provided by equation 8. It should be clear that it is necessary to use equation 6 for all R_j estimates when working with large cells having low input resistances, such as adult mammalian cardiac ventricular myocytes. When using larger cells, DWCR conditions can be improved by increasing R_{in} to mimic the properties of smaller cells. This can be accomplished by including pharmacological blockers of known background membrane currents (for example, K^+ currents) in the bath and pipette solutions of the preparation, as performed on adult guinea-pig ventricular cell pairs (18).

4.4 Effects of series resistance on junctional resistance measurements

Although alluded to already in the above discussions, the series resistance occurring primarily at the electrode tip figures prominently in the determination of R_j. Beginning with the ideal circuit, one would predict the I_j measurements to be underestimated by approximately 2% if R_{el} of the non-pulsed cell were to double from 10 to 20 MΩ. The effect of changing R_{el} on R_j measurements was examined by increasing R_{el2} to 20 MΩ. The pertinent current, voltage, and R_j measurements for this demonstration are listed in *Table 4*.

Patch-clamp analysis of gap junctional currents

Table 4. Analysis of electrode resistance changes on DWCR junctional resistance measurements.

	ΔV (mV)	ΔV_m [a] (mV)	V_j [b] (mV)	ΔI (pA)	I_j [c] (pA)	R_j [d,e] (MΩ)	R_j [e,f] (MΩ)
$R_{e12} = 20$ MΩ							
Cell 1	−40	−39.2	−38.5	−77	—	—	—
		{−39.2}	{−38.5}				
Cell 2	0	−0.7	—	36.8	−37.6	1087	1024
		{−0.7}				{+8.7%}	{+2.4%}
Cell 1	0	−0.3	—	36.8	−37.2	1087	1024
		{−0.4}				{+8.7%}	{+2.4%}
Cell 2	−40	−38.4	−38.0	−77.0	—	—	—
		{−38.5}	{−38.1}				
$R_{e12} = 100$ MΩ							
Cell 1	−40	−39.3	−36.0	−75.3	—	—	—
		{−39.2}	{−35.9}				
Cell 2	0	−3.1	—	33.3	−36.9	1201	971
		{−3.3}				{+20%}	{−2.8%}
Cell 1	0	−0.3	—	33.3	−33.6	1201	986
		{−0.3}				{+20%}	{−1.3%}
Cell 2	−40	−33.0	−32.6	−64.8	—	—	—
		{−33.5}	{−33.2}				

[a] Values in brackets calculated using equation 2.
[b] Values in brackets calculated using equation 4.
[c] Values in brackets calculated using equation 5.
[d] R_j calculation made using equation 7.
[e] Values in brackets are % error from specified R_j of 1000 MΩ.
[f] R_j calculation made using equation 6.

When ΔV of −40 mV was applied to cell 1, ΔI_1 and V_{m1} were not significantly altered from their previous measurements under more ideal conditions. ΔI_2 measured 36.8 pA and the predicted I_j was 37.6 pA. However, since R_{el} had doubled, ΔI_2 produced a −0.7 mV voltage drop across R_{el2}. This led to a further reduction in V_j to −38.5 mV and a conventional R_j estimate of 1087 MΩ, or an 8.7% over-estimation of R_j. Dividing −38.5 mV by −37.6 pA produced a corrected R_j estimate of 1024 MΩ. When V_2 is stepped to −40 mV, one would expect twice the previous voltage drop to occur across R_{el2}, while the accuracy of the $-\Delta I_1 = I_j$ estimate and V_{m1} should be unchanged from previous (ideal) results. Direct V_m and V_j measurements confirmed that V_{m2} was reduced by 1.6 mV, thereby producing an actual V_j of only −38 mV.

The conventional and corrected R_j estimates were again identical to the values obtained when V_1 was stepped.

When R_{el2} was further increased to 100 MΩ, the error in the conventional R_j estimates rose to 20%, while the corrected R_j estimates were within 3% of the specified R_j value. The primary causes for the 20% overestimation of R_j were that ΔI_2 underestimated I_j by 10% and V_{m2} produced an 8% drop in V_j when V_1 was stepped to −40 mV and V_{m2} amounted to only 84% of ΔV when ΔV was applied to V_2. The errors in the partner cell amounted to less than 3% of the total error in either case. Again, the resulting conventional R_j estimate was obtained independently of whether the low or high series resistance cell was pulsed. The observation of symmetrical R_j estimates makes detection of such systematic errors difficult unless R_{el} and R_{in} are independently measured for both cells.

4.5 Summary of effects of non-junctional resistance changes on R_j measurements

In the above demonstrations, the sources of error in making conventional R_j estimates were identified and the effects of changing non-junctional resistances on these parameters were examined while keeping $R_j = R_{in1} = 100 \times R_{el1}$. As one should comprehend by now, those sources of error include the voltage drop across each electrode produced by $\Delta I \times R_{el}$ which act synergistically to reduce V_j, and the underestimation of I_j by $-\Delta I$ of the non-pulsed cell ($-\Delta I_2$). All three parameters are readily calculated using equations 2 and 5, provided that ΔV_1, ΔI_1, ΔI_2, R_{el1}, R_{el2}, and R_{in2} are known. Because it is relevant to know how all of the various elements of the circuit are interacting to produce these errors, alternative expressions for determining I_j, ΔI_{m1}, and ΔI_{m2} have been derived in this section which emphasize these interactions. Rearranging equation 5, the $-\Delta I_2/I_j$ ratio can be simply expressed as $1/[1 + (R_{el2}/R_{in2})]$. Hence, $-\Delta I_2$ will always underestimate the true I_j. This point is illustrated graphically in *Figure 4A*. That fraction of I_j which is absent from ΔI_2 must be equal to ΔI_{m2} produced by the ΔV_1 pulse.

Assuming ΔV is always applied to cell 1 and the resulting I_j gives rise to ΔI_{m2}, it is useful to express ΔV_{m2} as a function of I_j. Given that $\Delta V_{m2} = \Delta I_{m2} \times R_{in2}$, it follows that:

$$\Delta V_{m2} = I_j \times R_{c2} \qquad [9]$$

where $R_{c2} = (R_{in2} \times R_{el2})/(R_{in2} + R_{el2})$. The above relationship can also be expressed as a function of ΔI_1 since $I_j = \Delta I_1 \times [R_{in1}/(R_{in1} + R_j + R_{c2})]$. The importance of keeping R_{c2} low, which is also dominated by R_{el2}/R_{in2}, is illustrated in *Figure 4B*. An R_{c2} of 0.01 GΩ corresponds to the example of $R_{in2} = 0.91$ GΩ and $R_{el2} = 0.01$ GΩ shown in *Table 2* and equation 9 accurately predicts the ΔV_{m2} of −0.4 mV for this example. $R_{c2} = 0.09$ GΩ matches the example shown in *Table 4* when $R_{in2} = 0.91$ GΩ and $R_{el2} = 0.10$ GΩ

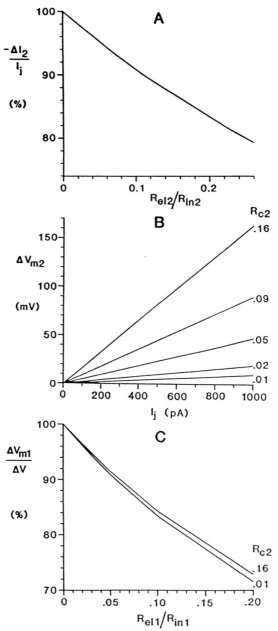

Figure 4. Effects of non-junctional membrane resistances on $\Delta I_2/I_j$, ΔV_{m2} and ΔV_{m1}. (**A**) Solid line depicts the percentage of I_j represented by $-\Delta I_2$ as a function of R_{el2}/R_{in2} ratio. The graph demonstrates that $-\Delta I_2$ routinely underestimates I_j. (**B**) The portion of I_j which does not appear in the $-\Delta I_2$ signal (= ΔI_{m2}) produces a ΔV_{m2} that changes in direct proportion to R_{c2}. (**C**) The percentage of ΔV_1 which manifests itself as ΔV_{m1} is influenced primarily by R_{el1}/R_{in1} when $R_j = R_{in1}$.

and again the predicted ΔV_{m2} of -3.0 mV agrees closely with the actual experimental value of -3.1 mV. So one can readily calculate I_j and ΔV_{m2} provided that ΔI_2, R_{el2}, and R_{in2} are known.

The accuracy of the $V_{m1} = \Delta V_1$ assumption can be conveniently expressed as the ratio of $\Delta V_{m1}/\Delta V_1$, or:

$$\Delta V_{m1}/\Delta V_1 = 1/\{1 + [(R_{el1}/R_{in1}) \times (R_{in1} + R_j + R_{c2})/(R_j + R_{c2})]\}. \quad [10]$$

In the above demonstrations, $R_{in1} = R_j = 1$ GΩ. Hence, equation 10 reduces to $1/\{1 + [(R_{el1}/R_{in1}) \times (2 + R_{c2})/(1 + R_{c2})]\}$. The effect of R_{el1}/R_{in1} on $\Delta V_{m1}/\Delta V_1$ for different R_{c2} values are presented in *Figure 4C*. In contrast to ΔV_{m2}, ΔV_{m1} is only slightly affected by a 16-fold increase in R_{c2} under these circumstances and $\Delta V_{m1}/\Delta V_1$ is approximately equal to $1/[1 + (2R_{el1}/R_{in1})]$. Given $R_{el1}/R_{in1} = 0.01$ and ΔV_1 of -40 mV, $\Delta V_{m1} \cong -39.2$ mV for the examples depicted in *Tables 2–4*. So it has been demonstrated that equation 10 provides an accurate prediction of ΔV_{m1} relative to ΔV_1 and $V_j = \Delta V_{m1} - \Delta V_{m2}$.

4.6 Effects of changes in junctional resistance

The demonstrations so far have dealt with the effects of changes in non-junctional resistances on the accuracy of R_j measurements obtained with two different equations (6 and 7), one which relies on two basic assumptions and a second which takes into account the errors incorporated into those assumptions. One question that remains to be examined is how changes in R_j affect the accuracy of these measurements, even when the R_{el} and R_{in} values approximate the relatively ideal conditions set forth in *Table 1*. To examine this question, the model circuit presented in *Table 1* was modified by replacing R_j with a 500 or a 100 MΩ resistor. The results of this demonstration are listed in *Table 5*.

Based on the expressions listed above, one would expect ΔV_{m1} and ΔV_{m2} to be altered by changes in R_j. When $R_j = 500$ MΩ, ΔI_2 was 73.5 pA and I_j was calculated to be -74.3 pA. Given that $R_{c2} = 0.01$ GΩ, equation 9 estimated $\Delta V_{m2} = -0.7$ mV, exactly as measured. So ΔV_{m2} was doubled owing to the nearly twofold increase in I_j. From equation 10, one would expect ΔV_{m1} to be 97% of ΔV_1 or -38.8 mV, which also agreed with the measured value. When $R_j = 100$ MΩ, ΔI_2 measured 312 pA and I_j was -315.4 pA. According to equation 9, $\Delta V_{m2} = -3.2$ mV and $\Delta V_{m1} = 0.91 \times \Delta V_1 = -36.4$ mV, which were within 0.1 mV of the measured values. Hence, the above expressions accurately predicted the errors incorporated into the conventional R_j estimates of 544 and 128 MΩ, which were 9% and 28% in excess of the specified R_j value. As before, the corrected R_j estimates were within 5% of the specified R_j value. Similar results were obtained when ΔV was applied to V_2, with the notable exception of the asymmetric ΔI_1 and ΔI_2 values which were more pronounced at lower R_j values.

Table 5. Analysis of junctional resistance changes on DWCR junctional resistance measurements

	ΔV (mV)	ΔV_m [a] (mV)	V_j [b] (mV)	ΔI (pA)	I_j [c] (pA)	R_j [d,e] (MΩ)	R_j [e,f] (MΩ)
$R_j = 500$ MΩ							
Cell 1	−40	−38.8	−38.1	−116	—	—	—
		{−38.8}	{−38.1}				
Cell 2	0	−0.7	—	73.5	−74.3	544	513
		{−0.7}				{+8.8%}	{+2.6%}
Cell 1	0	−0.6	—	75.3	−76.1	531	501
		{−0.8}				{+6.2%}	{+0.3%}
Cell 2	−40	−38.8	−38.1	−112.0	—	—	—
		{−38.9}	{−38.1}				
$R_j = 100$ MΩ							
Cell 1	−40	−36.3	−33.0	−368	—	—	—
		{−36.3}	{−33.2}				
Cell 2	0	−3.2	—	312	−315.4	128	105
		{−3.1}				{+27%}	{+5%}
Cell 1	0	−3.2	—	329	−332.3	122	100
		{−3.3}				{+20%}	{+0%}
Cell 2	−40	−36.3	−33.0	−357	—	—	—
		{−36.4}	{−33.1}				

[a] Values in brackets calculated using equation 2.
[b] Values in brackets calculated using equation 4.
[c] Values in brackets calculated using equation 5.
[d] R_j calculation made using equation 7.
[e] Values in brackets are % error from specified R_j of 1000 MΩ.
[f] R_j calculation made using equation 6.

It is apparent from equation 10 that as R_j decreases, R_{c2} will figure more prominently in determining the value of ΔV_{m1}. Furthermore, provided that R_{in2} is high, R_{c2} is approximately equal to R_{el2}. This is readily demonstrated by concomitantly increasing R_{c2} while lowering R_j. Under the conditions described above, an R_{el2} of 20 or 100 MΩ translates into an R_{c2} of 0.02 or 0.09 GΩ. When $R_j = 500$ MΩ and $\Delta V_1 = -40$ mV, V_{m1} increases from -38.8 mV to -39.0 mV when R_{c2} increases from 0.01 (*Table 5*) to 0.09 GΩ. When $R_j = 100$ MΩ, the same R_{c2} values led to an increase in V_{m1} from -36.3 to -37.6 mV. Calculations using equation 10 exactly matched these measured values. The effect of altering R_{c2} values on $\Delta V_{m1}/\Delta V_1$ at different R_j values is illustrated in *Figure 5*. For comparison, the $R_j/R_{in1} = 1$ example illustrated in

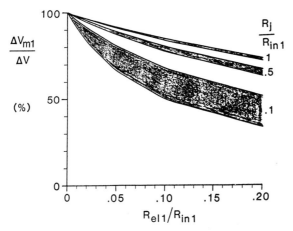

Figure 5. Effects of lowering junctional resistance on ΔV_{m1}. The shaded areas represent the range of $\Delta V_{m1}/\Delta V_1$ which occur for a given R_j/R_{in1} ratio as a function of R_{el1}/R_{in1}. The lower limit of each range is bounded by $R_{c2} = 0.01$ GΩ and the upper boundary is represented by $R_{c2} = 0.16$ GΩ.

Figure 4C is again presented in *Figure 5*. The shaded areas represent the range of $\Delta V_{m1}/\Delta V_1$ values obtained between $R_{c2} = 0.01$ and 0.16 GΩ for three different R_j/R_{in1} values and demonstrate the increased variability in $\Delta V_{m1}/\Delta V_1$ induced by R_{c2} as R_j is lowered.

While ΔV_{m1} will be increased by an increase in R_{el2} when $R_j < R_{in1}$, ΔV_{m2} and $-\Delta I_2/I_j$ will be seriously affected by the increase in R_{el2}. For example, when $R_j = 100$ MΩ, ΔV_{m2} increased from -3.3 mV at $R_{c2} = 0.01$ (*Table 5*) to -6.0 and -17.8 mV when $R_{c2} = 0.02$ and 0.09 GΩ. These combined changes in ΔV_{m1} and ΔV_{m2} produced a drop in V_j from -33.1 mV ($R_{c2} = 0.01$ GΩ) to -30.5 and -19.8 mV ($R_{c2} = 0.02$ and 0.09 GΩ, respectively). Under these same circumstances, $-\Delta I_2/I_j$ dropped from 0.99 at $R_{c2} = 0.01$ to 0.98 and 0.90 for $R_{c2} = 0.02$ and 0.09 GΩ. Hence, the error in the conventional R_j estimates increased to 38% and 131% for $R_j = 100$ MΩ and $R_{c2} = 0.02$ and 0.09 GΩ. Corrected R_j estimates were routinely within 4% of the specified R_j values.

4.7 Detection of gap junction channel currents

The resolution of single gap junction channel currents requires one additional condition to be placed on the DWCR preparation. The current noise must not exceed the amplitude of the channel currents. A high R_{in} value limits the non-junctional current noise and the amplifier noise should be reduced to its minimum value by proper shielding and grounding of the experimental apparatus, but R_j must also be high if channel activity is to be observed. As a general rule, gap junction channel activity can be resolved if $g_j = 1/R_j \leq 10 \times$

γ_j, where γ_j = single-channel conductance. Large errors in the V_j and I_j estimates are minimal under optimal channel recording conditions, so the major limiting factor is the resolution of the channel current amplitude(s). The variance of the channel current increases concomitantly with the number of open channels, which in part explains why a low number of open channels is required for adequate resolution of single-channel currents. Naturally, gap junction channels with larger γ_j, such as the 140–160 pS channel of embryonic chick heart, are more readily observed than lower γ_j channels (14). In all cases, it is advantageous to display at least some of the channel records as paired currents of equal magnitude and opposite polarity to convince the reader of their junctional origin. Large current fluctuations can also be produced by non-junctional sources, such as changes in R_s, but a majority of the additional current should be prevalent in the cell experiencing the change in R_{in} and any 'leak' current appearing in the partner cell will have the same polarity.

One approach to improving the probability of resolving channel currents in DWCR preparations is to pharmacologically increase R_j by partially uncoupling the cells with one of the better known uncoupling agents. The 1-alkanols, heptanol and octanol, and the volatile anaesthetic, halothane, have been used for this purpose (14, 18, 19). The use of these agents requires the assumption that they are acting to increase R_j by reducing the number of open channels without altering γ_j, a claim which is supported by channel open time and γ_j measurements (14, 19, 20). The presence of mutiple channels precludes a detailed kinetic analysis of channel gating and limits the usefulness of the information gained from the DWCR approach. The determination of γ_j and the ensemble averaging of multiple channel records to relate single channel activity to time-dependent changes in macroscopic I_j can be performed by the DWCR technique (15, 16). When reporting channel amplitudes, typically the value is reported for a given V_j, although a more complete description can be obtained by determining the single channel I–V (i_j–V_j) relationship for a given channel. If the i_j–V_j relationship proves to be linear, then γ_j equals the slope of the line. Methods for measuring channel amplitudes, estimating open state probabilities, and defining boundary conditions for channel event detection are presented in Section 7 of this chapter.

5. Double whole-cell vs. single patch approach

The gap junction channel has proved to be a particularly hard channel to gain access to in terms of monitoring its gating activity. All other channels are neatly placed in the plasma membranes of cells with one side of the their channel orfice facing the extracellular space and the other facing the cytoplasm. The gap junction channel is unique in that the orfices of the channel face the cytoplasms of two adjacent cells. Therefore the application of a single patch electrode to the external surface of a cell may result in the observation of many channel forms interfaced between external and internal environments but it will not permit the observation of the activity of intact gap junction channels.

The double voltage-clamp, an ancestor of the double whole-cell patch-clamp, was the first method used to monitor gap junctional membrane conductance exclusive of other membrane currents. The smallest current fluctuations observable with such systems were in the nanoampere range (1, 21). The introduction of patch-clamp methods allowed observation of current fluctuations in the picoampere range (2), a natural step was the application of patch-clamp methods to cell pairs. Thus double whole-cell patch-clamp methods were introduced (13, 22, 23). Unfortunately, the impedance of cell pairs was often low (less than 100 MΩ) and the number of gap junction channels high. These two factors, together or singly, often resulted in an inability to resolve unitary gap junction channel events. Under such conditions, the double whole-cell clamp yields results only slightly better than double voltage-clamp with microelectrodes. The only significant improvement being the potential to alter the internal mileau of the cell such that the electrolyte concentrations are relatively well established. This too has limitations (24).

Partial uncoupling of a cell pair with a high number of gap junction channels can result in the visualization of single-channel activity (22). But to observe the activity of single gap junction channels by the application of uncoupling agents compromises one's ability to study nascent activity. In fact, only when cellular impedance is high and gap junction channel density low is single-channel activity readily observable. This was the case in the study presented by Veenstra and DeHaan (23) where pairs of chicken embryonic heart cells were used.

The strengths of the double whole-cell patch method are an unambiguous view of gap junction membrane conductance and, when the number of active gap junction channels is low enough, observation of single channel activity. A weakness of this technique is the need for high cell impedance and hence small cell size, or in the case of large cells (working myocardium, see ref. 18), the use of external bathing media which blocks most or all of the plasma membrane channel conductance. This last approach is simply a way of making a large cell appear electrically as a small high impedance cell. The second weakness is a conditional one. If the number of intact functional gap junction channels is large it is difficult to observe single unitary activity with certainty. Individual gap junction channels often aggregate in groups of 100 or more. In some cases such as liver or heart there can be thousands of channels packed together in a single aggregation (25). Some preparations have gap junction channels grouped together in much smaller arrays such that only 10–50 channels are found in close association (26, 27). In many cases the number of aggregations of gap junctions is also high. A final problem is access resistance of the patch pipettes which can attain values as high as 20 or more MΩ.

Clearly, in cases where the junctional membrane resistance is low (1 µS, many channels) true voltage control of the junctional membrane is compromised.

As already noted, often cells are coupled to their neighbours via many gap junction channels, a condition which disallows easy observation of single gap junction channel activity. Monitoring single-channel activity from such preparations by directly patching the junctional membrane surface is an attractive alternative. The number of channels within the patch rather than the total number connecting two cells would be limiting in terms of the resolution of unitary events. The direct patch method allows for observation of a small population of gap junction channels from among many in differentiated cells. Furthermore, the ionic conditions under which one records can be more strictly controlled. The disadvantage of the method is that it requires extensive experimentation to demonstrate that the channels being recorded from behave as gap junction channels.

The following will outline methods which can be used to expose cytoplasmic surfaces and prepare them for patch recordings. In addition, analytical methods which allow for determination of the channel types and channel behaviour will be presented.

6. Achieving direct cytoplasmic access to gap junction channel

Figure 6 illustrates one scheme for exposing the interior of cell membranes. The approach takes advantage of the fact that some cells form gap junctions with other cells over extensive regions of close contact. Dissociation of cells (pulling cells apart) can result in many cells having on their surfaces intact gap junction membranes (27, 28, 29). The procedures for isolating single or small groups (pairs) of myocardial cells are well established (see ref. 13). These dissociation procedures normally allow for healing over of cells, a process of many hours. Healing over appears to be made necessary because of the fact that gap junctions are often found intact on isolated cells. Often the associated membranes of such an exposed gap junction form a vesicle with time when bathed in normal saline (Ca^{2+} = 1 or more millimolar; ref. 27). The time-course of vesicle formation and the effect of Ca^{2+} are unknown. No attempt to date has been made to recorded gap junction channel activity from such a preparation.

With regard to large cells such as the septate axons of the earthworm or crayfish, the axonal processes of neurones form long cylindrical axons which abut end to end. The points of contact at the end-to-end interfaces contain gap junctions (26). By simply cutting the axon perpendicular to the long axis of the axon near an interface (end-to-end abutment) the cytoplasmic membrane face of the cut axon is exposed to the extracellular media usually within

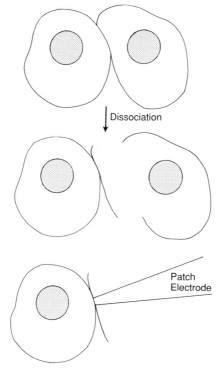

Figure 6. Preparation of junctional membrane. Illustration of cell dissociation in which one of two cells retains an intact gap junction membrane. The exposed membrane can then be patched.

easy reach of a patch electrode. This procedure has been successfully used by Brink and Fan (30), using the earthworm septate axons. Seals of 500 MΩ to many GΩ can be attained routinely. Usually, only slight suction was required to attain seals in the study of Brink and Fan (30).

6.1 Optimal ionic conditions

A critical step is the cleaning of the cytoplasmic face of 'ruptured' or dissociated cells. This cleaning is necessary to ensure adequate seals between the patch electrode and the membrane. The best approach to conditioning a cytoplasmic face is the one adopted by Bezanilla (31). To patch the cytoplasmic surface of squid axon Bezanilla, once having sliced open the axon, irrigated the axon interior with sea water. The sea water contained among other ions about 20 mM $CaCl_2$. Mixing of the sea water with axoplasm enabled activation of proteases activated by high Ca^{2+}. Exogenous application of enzyme was not as successful as the Ca^{2+} treatment. A similar result is obtained in the septate axons by making a cut near an interface between two axons where the bathing media is normal saline. In this case the Ca^{2+} concentration need only be 1 mM. The exposure time to the saline is critical to the formation of good seals. Recall that most gap junction channels while permeable to Ca^{2+} in low concentration are blocked by it at higher concentrations. Further, as is the case with the activation of most proteolytic processes, exposure to prolonged action can cause irreversible damage (loss of channel activity being one of many). As a general rule for each system in which this procedure is tried the length of Ca^{2+} exposure necessary to improve seals, but not result in loss of activity, will vary. In the case of the earthworm the length of time for Ca^{2+} exposure was on the order of 5–15 min at most.

For dissociated cells, the cleansing of exposed cytoplasmic surfaces becomes a natural part of the procedure to isolate cells in that the cells as they are dissociated are exposed to millimolar levels of Ca^{2+} normally.

6.2 Avoiding intracellular membranes

In the squid axon (intact not sliced open) it has been found that small cuts into the axon result in the formation of large membrane blebs (32). The blebs are thought to be forming as a result of recruitment of intracellular membranes to the surface membrane. This process requires millimolar levels of extracellular Ca^{2+}. As an exercise, lower levels of Ca^{2+} should be used during the exposure procedure to ensure that bleb formation from internal membranes is not occurring. Ca^{2+} concentrations as low as 0.1 μM have worked on the earthworm preparation.

6.3 How to know if one has gap junction channels in a patch

The real answer to this question is: one cannot know absolutely. Although, comparison with known gap junction channel function and blockade of other known channel types constitute rather stringent criteria for determining if a channel type is a gap junction channel.

Experimental procedures for determining channel type:

(a) Use specific channel blockers within the pipette and bathing media.
(b) Test the selectivity or lack of selectivity of the channel types found.
(c) Test the activity of observed channel types in the cell-attached mode and detached modes.
(d) Test the effects of known gap junction channel blockers on any suspect channel (Ca^{2+}, H^+, octanol).
(e) Test for symmetric voltage-dependence or voltage-independence.
(f) Comparison of macroscopic behaviour with channel activity.

These six criteria are the basic approaches one can use to determine if a channel type is gap junction-like.

6.3.1 Blockers

One of the most powerful discriminators is the use of channel blockers. For example the use of CsCl as the major cation/anion source eliminates many K^+ channels. Along the same lines tetraethylammonium chloride (TEA) has the same purpose. A typical electrolyte solution from the pipette and/or bath when using the direct patch approach is listed in *Table 6*. This saline contains electrolytes that should be carried by a large-bore channel. In addition, the cations present have specific abilities with regard to blocking other channel types. Both Cs^+ and TEA block K^+ channel activity. Co^{2+} and Ni^{2+} block Ca^{2+} channels. Many Ca^{2+} channels are able to allow the passage of monovalent cations when the free Ca^{2+} concentration is low. Often the single-channel conductances are in the 40–100 pS range (33). Cd^{2+} is another all purpose Ca^{2+} channel blocker. The free Ca^{2+} concentration for the saline shown above is in the 10–100 nM range. Raising the free Ca^{2+} level to 0.1–1 mM should alter or block the activity of any gap junction channel (34). If one is concerned that a channel is an anion channel (Cl^- for example) the presence of Zn^{2+} should preclude any significant activity. This is one example, it is possible to add many other agents such as aminopyridine (1–4 mM) which will block the activity of many K^+ channels.

Table 6. Composition of electrolyte solution for direct patch technique

Ingredient	Concentration (mM)	Ingredient	Concentration (mM)
CsCl	135	$CaCl_2$	0.1
TEACl	30	$CoCl_2$	1
KCl	1	$NiCl_2$	1
NaCl	1	$ZnCl_2$	0.5–2
Hepes	10	TTX	0.01–0.001 (optional)
EGTA	0.6	pH 7.0	

The use of blocking agents assumes most all channel types are known, and while it might be comforting to find a channel that is relatively unaffected by known channel blockers, such an ability is not necessarily unique to gap junction channels nor expected. In conjunction with other criteria though, it is a very powerful tool in determining channel type.

6.3.2 Selectivity

To determine the selectivity or lack of selectivity of a channel it is best to have identical salines on either side of a patched membrane. This translates into the detached mode (inside-out when considering single membrane channels) where the bath saline and pipette saline are experimentally defined. This experimental configuration is a control condition as the equilibrium potentials for all ions is 0 mV. Once having established that channel activity is present under such conditions it is possible to alter the contents of the bath saline such that the equilibrium potentials for specific ions is changed. By plotting the *I–V* curve for the unitary current vs. holding potential before and after such a bath exchange the ion selectivity of the channel can be determined (35). One needs to establish whether or not a channel is a cation channel or able to pass cations and anions. For gap junctions the ability to pass both cations and anions is a must, as many negatively changed fluorescent probes of 1.0 nm dimension are able to diffuse through gap junctions (36, 37). *Figure 7* shows a semi-log plot of the conductivity ratios in dilute solution (relative to K^+) for Na^+, Cs^+, TMA^+, and Cl^- as well as the conductance ratios determined for the gap junction channel of earthworm using the approach described above. The conductance ratios for one of the porin channels is also given. It is a well documented channel type. Porin OmpF is one porin type able to pass cations and anions, and has an estimated channel diameter of 0.9–1.0 nm. Since both porin and the gap junction channel are relatively non-selective it is possible to estimate the diameter of the gap junction channel. In this case the data

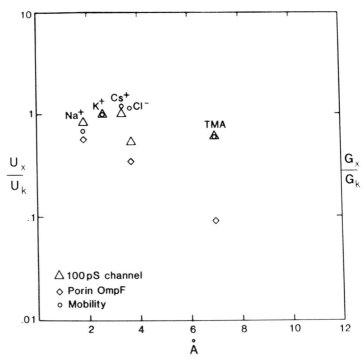

Figure 7. Relative mobilities and conductances of various molecules. Semilog plot of the ratio of mobilities (*circles*), ratio of conductances for the Porin OmpF channel (*diamonds*) and ratio of conductances for the earthworm gap junction channel (*triangles*). K^+ mobility or conductance was the denominator in calculating the ratios.

indicated that the gap junction channel might well have a larger bore than porin. Application of diffusion limiting channel models (38) allows for more precise estimation.

The lack of selectivity for the gap junction channel implies that the channel is relativistically 'perceived' by the ions to be similar to free solution. The selectivity sequences for the gap junction channel and the porin are similar to an Eisenmann I or II sequence (17).

6.3.3 Use of attached and detached recording modes

When identifying any channel it is important to determine its behaviour in the attached mode and detached mode. Often the activity of a channel will vary from one mode to the other. If for no other reason comparison can reveal whether silent patches in the detached mode are due to lack of channels or the need for some cytoplasmic element. The opposite might also be true.

6.3.4 Test the effects of known blockers

Once channel activity has been established and something of the selectivity is known it is paramount to establish whether the application of gap junction channel blockers to the bath saline can alter the activity of the channels being recorded. Recall that the pipette and bath salines normally contain free Ca^{2+} concentrations of 100 nM or less. Raising the free Ca^{2+} to 100 µM or 1 mM in the bath is equivalent to cellular processes which close gap junction channels. Likewise, increasing the H^+ concentration in the bath should also result in a marked decrease in channel activity. Few channels respond to elevations of Ca^{2+} in the form of increased closed time. In fact, most are activated by the presence of Ca^{2+}. The maxi-K channel is one example (39). Thus the Ca^{2+} control is useful. Unfortunately, elevation of H^+ can affect many systems (many channel types) so it is not the best single blocker to use to close gap junction channels. Other easily applied gap junction channel blockers are octanol and similar compounds (40). The only issue with agents like octanol is difficulty with knowing the concentration needed to alter activity and often the need for a second vehicle to get any of the agent (octanol) into solution. Octanol and its derivatives are also potentially able to affect other channels within a cell. If a channel is affected by gap junction uncoupling agents but unaffected by other blocking agents (TEA, Co^{2+}) one has gone a long way in demonstrating that the channel being observed is indeed a gap junction channel.

The direct patch method allows only the bath side of the patch to experience ionic changes and is also the only surface exposed to uncoupling agents. Often in double whole-cell patch or multicellular experiments agents are administered such that both sides of a junction are exposed. The early experiments performed by Loewenstein (34) did demonstrate that only one side of a junction need be exposed to an uncoupling agent to result in an increase in junctional membrane resistance (closure of channels).

6.3.5 Test for voltage dependence

Having established channel activity, one natural parameter to alter is the voltage held across the patch. A number of investigators have shown, using the double whole-cell patch method, that many gap junctions in embryonic systems and early developmental stages display varying degrees of voltage dependence (15). Voltage-dependence for gap junction channels is characterized by decreasing channel activity with increasing voltage across the channel. The importance of monitoring channel activity relative to voltage-dependence is best appreciated when considering macroscopic measurements. For there are many examples of gap junctional membranes which show no voltage-dependence (41, 42) and many which show varying degrees of voltage-dependence (15).

6.3.6 Macroscopic behaviour vs. single-channel activity

Macroscopic measurements are the most common form in which gap junction channel activity has been displayed. The following questions are examples which demonstrate the utility of comparison of single channels and macroscopic measurements. Do the single channels behave in a similar fashion as the whole junctional membrane when exposed to similar agents? Does pH have the same effect for example? If voltage-dependence is observed in the macroscopic records is the same true for the single-channel data? Does octanol have similar effects on both macroscopic and microscopic behaviours? The exact nature of the comparisons is highly dependent on the particulars of the macroscopic behaviour. For time-dependent behaviour, the averaging of single or even multichannel recordings should produce records similar to macroscopic time-dependent changes in current, for example (15).

7. Channel analysis: sampling methods

An important issue with regard to recording channel activity is the form of the sampling. In general, when recording data via an A/D converter, the sampling rate should be many times faster than the low bandpass filter settings (43). A related question is how large a sample is adequate to represent the behaviour of a channel or channels within a patch. If a channel opens/closes 100 times in a minute on average, then to observe 5000 events would require at least 50 min of recording. Because channel activity is subject to rapid alteration, such as mode shifting (39), rather large sample sizes are needed to determine the causal relationships between applied agents and channel open time for example. Obviously, agents which result in all or no like responses need not be considered, but when more subtle effects are observed caution is required. It is best to rely on analytical approaches to discern trends in data (44, 39).

R. D. Veenstra and P. R. Brink

7.1 Simple analytical approaches

Figure 8 illustrate data from the earthworm preparation. The multichannel recording was obtained via direct patch, and the data was sampled at 360 μsec and was filtered during data acquisition via computer at 200 Hz. This particular recording is 7.2 sec long. Note at the right-hand margin the vertical histogram for the data displayed.

Figure 9 shows the amplitude histogram from a multichannel patch. Note the noise about each peak broadens with more and more channel activity. This is a common feature of multichannel records and often it is the open channel variance in combination with the amplifier noise which limits the ability to resolve unitary events. The amplitude histogram of *Figure 9* is from a data set which was 43.2 sec long. Again, filtration of the raw data during acquisition was 200 Hz and the sampling rate was 360 μsec. This form of histogram is a point to point histogram and does not represent an event histogram. The vertical axis is the number of points, throughout the entire acquired record, at each current level (resolution to 0.1 pA). To count events

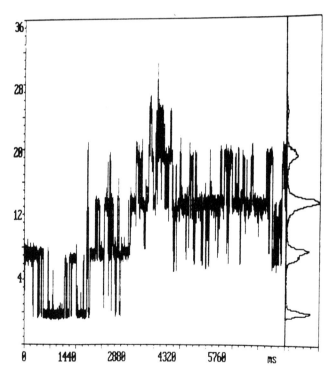

Figure 8. Channel currents of earthworm junctional membrane. This multichannel recording was 7.2 sec long. The extreme right-hand margin contains a histogram of the 7.2-sec data sample. Detached patch, holding potential = +65 mV, bath and pipette contained CsCl saline (low Ca^{2+}), Y-axis is in pA.

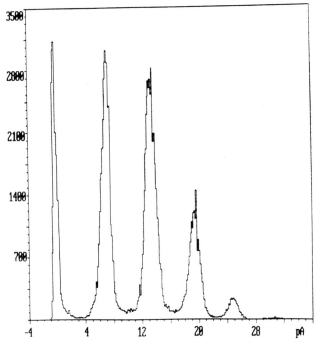

Figure 9. Channel amplitude histogram. Amplitude histogram of the current distribution for 43.2 sec of data, the data of *Figure 8* is a segment of *Figure 9*. Note that the noise (variance) about the open peaks (5) increases as more channels are open. Vertical axis is the number of sampled points at different current levels.

one need only place discriminators at intervening levels between open states and ask for the total number of transitions from one level to another. *Figure 10* shows the data of *Figure 8* with discriminators in place. The number of events depicted in *Figure 10* are a subset for the entire histogram (*Figure 9*) where the total number of events (transitions) was 1233. *Table 7* summarizes the events from level to level (discriminators as shown in *Figure 10*).

Table 7. Summary of channel event detection

Transition level	No. of events
level 1	87
level 2	333
level 3	467
level 4	277
level 5	63
level 6	6

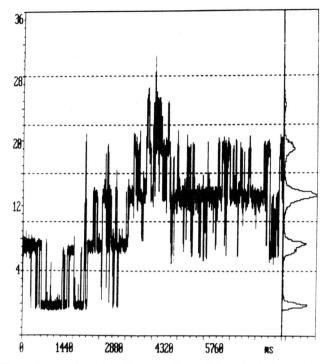

Figure 10. Channel event detection. Same as *Figure 8*, but discriminators (*dashed lines*) have been placed on the record to show how events can be counted.

The use of discriminators requires some care. If a channel is prone to substate then it is possible for a discriminator to traverse through an open substate. This will cause the event counting to be skewed toward the sampling rate. Diligent monitoring of the acquired data is the best way to avoid this sort of problem.

References

1. Spray, D. C., Harris, A. L., and Bennett, M. V. L. (1981). *J. Gen. Physiol.*, **77**, 77.
2. Hamill, O. P., Marty, A., Neher, E., Sakmann, B., and Sigworth, F. J. (1981). *Pflügers Arch.*, **391**, 85.
3. Veenstra, R. D. (1991). *J. Cardiovasc. Electrophysiol.*, **2**, 168.
4. Spector, I. (1983). In *Single Channel Recording* (ed. B. Sakmann and E. Neher), pp. 77-90. Plenum Press, New York and London.
5. Trube, G. (1983). In *Single Channel Recording* (ed. B. Sakmann and E. Neher), pp. 69-76. Plenum Press, New York and London.
6. White, R. L., Spray, D. C., Campos de Carvalho, A. C., Wittenberg, B. A., and Bennett, M. V. L. (1985). *Am. J. Physiol. (Cell Physiol. 18)*, **249**, C447.

7. Rook, M. B., Jongsma, H. J., and van Ginneken, A. C. G. (1988). *Am. J. Physiol. (Heart Circ. Physiol. 24)*, **255,** H770.
8. Marty, A. and Neher, E. (1983). In *Single Channel Recording* (ed. B. Sakmann and E. Neher), pp. 107–22. Plenum Press, New York and London.
9. Fischmeister, R., Ayer, R. K., and DeHaan, R. L. (1986). *Pflügers Arch.,* **406,** 73.
10. Corey, D. P. and Stevens, C. F. (1983). In *Single Channel Recording* (ed. B. Sakmann and E. Neher), pp. 53–68. Plenum Press, New York and London.
11. Sigworth, F. J. (1983). In *Single Channel Recording* (ed. B. Sakmann and E. Neher), pp. 3–35. Plenum Press, New York and London.
12. Giaume, C. (1991). In *Biophysics of Gap Junction Channels* (ed. C. Peracchia), pp. 175–90. CRC Press, Boca Raton, Florida.
13. Weingart, R. (1986). *J. Physiol., Lond.,* **370,** 267.
14. Veenstra, R. D. and DeHaan, R. L. (1988). *Am. J. Physiol. (Heart Circ. Physiol. 19),* **250,** H453.
15. Veenstra, R. D. (1990). *Am. J. Physiol. (Cell Physiol. 27),* **258,** C447.
16. Veenstra, R. D. (1991). *J. Membr. Biol.,* **119,** 253.
17. Jack, J. J. B., Noble, D., and Tsien, R. W. (ed.) (1983). In *Electric Current Flow in Excitable Cells*, pp. 225–6. Clarendon Press, Oxford.
18. Rudisuli, A. and Weingart, R. (1989). *Pflügers Arch.,* **415,** 12.
19. Burt, J. M. and Spray, D. C. (1988). *Circ. Res.,* **65,** 829.
20. Veenstra, R. D. and DeHaan, R. L. (1989). In *Cell Interactions and Gap Junctions* (ed. N. Sperelakis and W. C. Cole), Vol. II, pp. 65–83. CRC Press, Boca Raton, Florida.
21. Brink, P. R., Mathias, R. T., Jaslove, S. W., and Baldo, G. (1988). *Biophys. J.,* **53,** 795.
22. Neyton, J. and Trautmann, A. (1985). Single-channel currents of an intercellular junction. *Nature, London,* **317,** 331.
23. Veenstra, R. D. and DeHaan, R. L. (1986). *Science,* **233,** 972.
24. Mathias, R. T., Cohen, I. S., and Oliva, C. (1990). *Biophys. J.,* **58,** 759.
25. Goodenough, D. A. (1975). *Cold Spring Harbor Symp. Quant. Biol.,* **40,** 37.
26. Kensler, R. W., Brink, P. R., and Dewey, M. M. (1979). *J. Neurocytol.,* **8,** 565.
27. Mazet, F., Wittenberg, B. A., and Spray, D. C. (1985). *Circ. Res.,* **56,** 195.
28. Barr, L., Dewey, M. M., and Berger, W. (1965). *J. Gen. Physiol.,* **48,** 797.
29. Barr, L., Berger, W., and Dewey, M. M. (1968). *J. Gen. Physiol.,* **51,** 346.
30. Brink, P. R. and Fan, S. F. (1989). *Biophys. J.,* **56,** 579.
31. Bezanilla, F. (1987). *Biophys. J.,* **52,** 1087.
32. Stein, P. G., Fishman, H. M., and Tewari, B. J. (1989). *Biophys. J.,* **55,** 588a.
33. Wray, D. W., Norman, R. I., and Hess, P. (ed.) (1989). *Annals N.Y. Acad. Sci.,* **560,** 1.
34. Loewenstein, W. R. (1981). *Physiol. Rev.,* **61,** 829.
35. Hille, B. (1984). *Ionic Channels of Excitable Membranes.* Sinauer Associates, Sunderland, Massachusetts.
36. Burt, J. M. (1987). *Am. J. Physiol. (Cell Physiol. 22),* **253,** C607.
37. Brink, P. R. and Ramanan, S. V. (1985). *Biophys. J.,* **48,** 299.
38. Lauger, P. (1979). In *Membrane Transport Processes* (ed. C. Stevens and R. Tsien), pp. 17–28. Raven Press, New York.
39. McManus, O. B. and Magleby, K. L. (1988). *J. Physiol., Lond.,* **402,** 79.

40. Burt, J. M. (1991). In *Biophysics of Gap Junction Channels* (ed. C. Peracchia), pp. 75–96. CRC Press, Boca Raton, Florida.
41. Johnson, M. F. and Ramon, F. (1982). *Biophys. J.*, **39**, 115.
42. Verselis, V. and Brink, P. R. (1900). *Biophys. J.*, **45**, 147.
43. Colquhoun, D. and Sigworth, F. J. (1983). In *Single Channel Recording* (ed. B. Sakmann and E. Neher), pp. 196–7. Plenum Press, New York and London.
44. Ramanan, S. V. and Brink, P. R. (1990). *Biophys. J.*, **57**, 893.

9

Biochemical approaches for analysing *de novo* assembly of epithelial junctional components

MANIJEH PASDAR

1. Introduction

The epithelial intercellular junctional complex is composed of four major components: tight junctions (1–4), adherens junctions (5–8), desmosomes (1, 9–12), and cell adhesion molecules (13–22). These components rapidly appear on the plasma membrane of adjacent cells upon induction of cell–cell contact. A common feature of these junctional components is their apparent structural similarity (*Figure 1A*). Each junctional component consists of an extracellular or membrane core domain and an intracellular or cytoplasmic plaque domain to which cytoskeletal elements are attached. In all identified cases the extracellular domain consists of glycosylated integral membrane proteins and the cytoplasmic domain comprises non-glycosylated proteins (*Figure 1B*). These integral membrane and cytoplasmic proteins presumably go through different biosynthetic pathways (*Figure 2*). The glycoproteins of the membrane core domain are synthesized and core glycosylated in the endoplasmic reticulum. These proteins are then transported to the Golgi complex, where they become complex glycosylated. Subsequently, the glycoproteins are transported via transport vesicles to the plasma membrane. The protein components of the cytoplasmic domain however, are synthesized on free ribosomes and are then transported to the plasma membrane. Consequently, the assembly of a multi-subunit junctional complex requires the coordinated expression and processing of the constituent proteins and their final association in the fully assembled structure. Therefore, a comparative analysis of the biosynthetic pathways of the protein constituents of a multi-subunit complex should provide valuable insights into mechanisms regulating its assembly. These types of analysis can easily be performed using metabolic labelling in pulse-chase studies in conjunction with various inhibitors of glycoprotein processing and transport. The purpose of this chapter is to describe the methodology used in analysing the mechanisms regulating assembly of a

Biochemical approaches for analysing de novo assembly

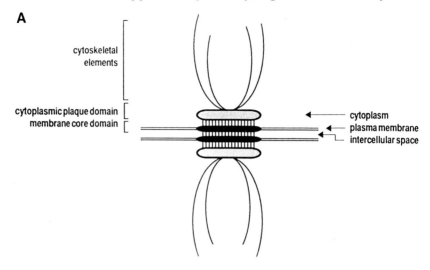

Figure 1. Components of the epithelial intercellular junctional complex. (A) A generic diagram representing the structural domains of any component of the intercellular junctional complex. (B) Protein composition of the structural domains of each specific components of the intercellular junctional complex.

multi-subunit junctional complex using desmosomes as a model system (see also Chapter 10).

Desmosomes are probably the most well characterized component of the intercellular junctional complex (refs 11–18; see also Chapter 6). The membrane core glycoproteins of desmosomes, known as desmogleins (DG), include DGI (M_r 150 000), DGII/DGIII (also known as desmocollins I/II); M_rs 120 000/110 000), and DGIV (M_r 22 000). The desmosomal cytoplasmic

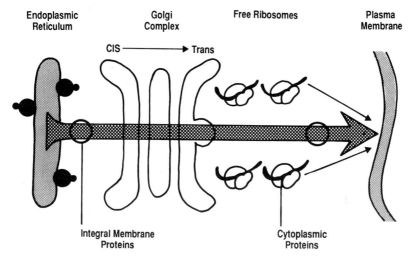

Figure 2. Schematic diagram of the biosynthetic pathways of a membrane core and a cytoplasmic protein constituents of a multi-subunit junctional complex.

plaque domain consists of non-glycosylated proteins called desmoplakins (DP): DPI (M_r 250 000), DPII (M_r 210 000), DPIII (also known as plakoglobin; M_r 83 000) and DPIV (M_r 78 000). With the exception of DPIII, all of the desmosomal proteins are specific to this structure. Desmosomes are present in abundance in established epithelial cell lines in culture and like other components of the junctional complex their assembly is strictly cell–cell contact-dependent (13, 19, 23, 24). Therefore, by modulating cell–cell contact through adjustment of the Ca^{2+} concentration in the growth media desmosome assembly can be synchronously regulated in culture, and the kinetics of synthesis, processing, and stability of the constituent proteins can be analysed under different conditions. Another property of desmosomes is that they can be easily identified in cultures by immunofluorescence and electron microscopy (EM). As a result, a combination of biochemical and morphological analyses of desmosome assembly in (Madin–Darby canine kidney) (MDCK) epithelial cells (25) has provided valuable information on the assembly process of this type of junction.

2. Establishment of cultures for *de novo* asssembly of desmosomes

A prerequisite to analyse *de novo* assembly of desmosomes is the absence of any old desmosomal components (24, 26–28) at the time of induction of cell–cell contact and assembly of new junctions. Since desmosome stability is cell–cell contact–dependent, maintenance of cultures of MDCK cells under

Biochemical approaches for analysing de novo *assembly*

conditions in which no cell–cell contact is present ensures the complete degradation of any proteins remaining from old desmosomes. To achieve this, cultures of MDCK cells are passaged through a low calcium pre-culture step.

Protocol 1. Low calcium pre-culturing

The composition of solutions and culture media are listed in the appendix at the end of this chapter.

1. Trypsinize cultures of MDCK cells at about 70–75% confluency and re-plate in medium containing normal Ca^{2+} concentration (1.8 mM, HCM) at single-cell density $1–1.5 \times 10^4$ cell/cm^2. Incubate 24–30 h.
2. Re-trypsinize cultures, plate cells at the same single-cell density in same medium and incubate another 24–30 h.
3. Trypsinize these cultures and plate cells at confluent density $1–1.5 \times 10^5$ cell/cm^2 (see *Protocol 3*) in media with low Ca^{2+} concentration (5 μM, LCM). 12 to 24 h later, cultures can be used for kinetic studies.

Under the conditions described in *Protocol 1*, cultures are maintained in the absence of cell–cell contact at least 72 h and desmosomal proteins that exist at the beginning of the pre-culture step degrade (turnover rate of various desmosomal components in the absence of cell–cell contact is 2–8 h). These are called 'contact naïve' cells. EM studies of cultures maintained under these conditions have confirmed that there is little or no cell–cell contact between closely apposed cells and there is no evidence of assembly of desmosomes or other components of epithelial junctional complex (29).

Replacement of LCM with HCM results in synchronous induction of cell–cell contact and initiates desmosome assembly within 15 min (26). Cultures prepared under these conditions can be used for kinetic analysis of synthesis, processing, transport, and stability of the protein constituents of a junctional component under different conditions.

3. De novo assembly studies

3.1 General techniques

Protocol 2. Preparation of collagen

Collagen (Type VII—acid soluble) can be prepared from rat tail (30, 31) or purchased commercially (Sigma). To prepare:

1. Store freshly obtained tails from 6-month-old rats in the freezer until convenient to use.

2. Place 5–10 tails in 95% ethanol to thaw.
3. Beginning at the distal tip, fracture the tail successively into smaller pieces, using haemostats.
4. Twist and pull each piece of tendon from the remainder of the tail. Cut the long silvery tendon strands and drop them into a Petri dish containing distilled water.
5. With two fine forceps tease tendons into thin strands and place them in 500 ml of acetic acid diluted 1/1000 in water on a stirrer.
6. Incubate at 4°C overnight while stirring.
7. Transfer into 50-ml centrifuge tubes and spin at 26 000 g for 2 h.
8. Remove the supernatant, a transparent jelly-like solution, and discard the remaining solid residue.
9. Store supernatant in refrigerator. If required, dilute the collagen as needed with 1/1000 acetic acid.

Protocol 3. Preparation of cultures

1. Coat 35 mm tissue culture Petri dishes with collagen prior to plating cells. Pour 2–3 ml of collagen solution into a dish and decant it thoroughly into the next dish or back into the container. Do not aspirate or pipette the excess collagen.
2. Sterilize and dry collagen-coated dishes by placing them under the UV light in laminar flow hood for 2–5 h. Cover collagen-coated Petri dishes with their tops, and store under sterile conditions at room temperature (RT). Collagen-coated dishes can be stored for several weeks.
3. Use pre-cultured MDCK cells (see *Protocol 1*) and plate $2-2.5 \times 10^5$ cells/cm^2 (confluent density) on collagen-coated 35-mm Petri dishes in LCM.
4. Change the media after 6 h with fresh LCM. Incubate 12–24 h and proceed with experiment.

Protocol 4. Metabolic labelling

1. Rinse cells maintained in LCM twice with LCM-MET, HDF, or DPBS.
2. Add 2.5 ml LCM-MET and incubate at 37°C 30 min.
3. Aspirate LCM-MET.
4. Label each 35 mm Petri dish with 125 μCi [^{35}S]methionine (translation grade) in 500 μl of LCM-MET and incubate for 15 min at 37°C. (The amount of radioactive precursor used depends on several factors, including the length of the labelling period, the relative abundance and the

Protocol 4. *Continued*

methionine content of the protein. Translabel (ICN), a combination of [^{35}S]methionine and [^{35}S]cysteine, can also be used.)

5. Aspirate labelling medium and rinse the cells 2–3 times with HDF.
6. For turnover studies add LCM- or HCM-chase medium containing >10 000-fold excess cold methionine to the cultures and continue incubation at 37 °C.

Protocol 5. Extraction and cell fractionation

This procedure is performed either on ice or at 4 °C.

1. Rinse the labelled cultures three times with ice-cold Tris–saline/PMSF.
2. Add 1 ml of extraction buffer to each 35 mm Petri dish. Usually this buffer contains non-ionic detergent such as Nonidet P-40 (NP-40) or Triton X-100 and various concentrations of salt. The extraction buffer is used to separate 'assembled' from 'non-assembled' junctional proteins. Cytoskeleton extraction buffer (CSK; ref. 32) is used to separate insoluble desmosomal proteins (presumably associated with cytoskeleton) from the soluble proteins.
3. Extract 10 min on a rocker platform.
4. 10 min into extraction with CSK buffer add 100 μl of 2.5 M $(NH_4)_2SO_4$ (to a final concentration of 250 mM) and continue extraction for an additional 5 min.
5. Remove the dishes from rocker, scrape cells up with a rubber policeman, and transfer to 1.5-ml tubes.
6. Spin at 48 000 g for 10 min.
7. Decant the supernatant into a new tube and quick-freeze it in liquid nitrogen.
8. Add 100 μl of SDS-immunoprecipitation buffer to the pellet, vortex, and heat in boiling water for 5 min.
9. Bring the volume up to 1 ml by adding 900 μl of CSK buffer and freeze it in liquid nitrogen. Samples can be processed for immunoprecipitation immediately or within a few days. Do not store the samples for long periods of time, as this will decrease the intensity of the signal due to the decay of [^{35}S] methionine ($t_{1/2}$ = 88 days).

Protocol 6. Immunoprecipitation

1. Quickly thaw the frozen samples in a 37 °C water bath.
2. Pre-clear samples by addition of 10–20 μl pre-immune serum of the animal in which the antibodies are raised and 25–35 μl of *Staphylococcus*

aureus (Calbiochem) cells. If monoclonal antibodies are used preclearing may not be required. *Staphylococcus aureus* cells are commercially available in 10% suspension in PBS containing sodium azide. For use in pre-clearing, *S. aureus* cells are washed twice in high-stringency immunoprecipitation buffer and resuspended in the same buffer in the original volume.

3. Incubate at 4°C for 30–60 min.
4. Spin in microfuge (14 000 g) for 5 min at 4°C. Decant the supernatant into clean tubes.
5. Add primary antibodies. The concentration of the antibodies depends on the titre of each specific antibody. For most polyclonals and ascites fluids 5–20 µl (in 1 ml of cell extract; final concentration of 1/50–1/200) is usually sufficient. For hybridoma culture supernatant 50–100 µl of an ammonium sulfate-cut supernatant is sufficient. In either case antibodies must be titrated and the optimum concentration determined.
6. Incubate at 4°C for 1–1.5 h. The optimum incubation time for primary antibodies should be determined for each individual antibody; unnecessarily long incubations result in higher background.
7. Add 25–35 µl protein A Sepharose CL-4B beads (Pharmacia) to each tube and incubate for 1.5–3 h at 4°C on a rocker or rotator.
8. Spin in microfuge (14 000 g) for at least 1 min at 4°C.
9. Aspirate the supernatant carefully without removing the beads. This can be done by attaching a fine-tip transfer pipette to a Pasteur pipette hooked to an aspirator. Wash the beads as follows:
10. Add 650 µl HS-buffer to each tube and vortex, then under layer 150 µl HS-buffer + sucrose.
11. Spin 1 min in microfuge at 4°C and aspirate supernatant.
12. Add 800 µl of high-salt immunoprecipitation wash buffer and vortex.
13. Spin 1 min in microfuge at 4°C and aspirate supernatant.
14. Add 800 µl low-salt immunoprecipitation wash buffer and vortex.
15. Spin 1 min in microfuge at 4°C and aspirate supernatant. The number of washes and the buffers used depends on the proteins under study. For non-glycosylated proteins (for example, desmoplakins) two HS ± sucrose washes followed by 1 low-salt wash seems sufficient. For glycoprotein (for example, DGs) 1 HS ± sucrose wash is followed by 1 high-salt and 1 low-salt washes.
16. Resuspend the final pellet in 45 µl of 1 × sodium dodecyl sulfate (SDS) sample buffer, vortex, and heat in boiling water for 5 min.
17. Load on to the appropriate SDS-polyacrylamide gels.

Quantitative immunoprecipitation is essential in comparing the kinetics of synthesis, processing, and stability of different proteins. It requires an efficient antigen–antibody binding as well as efficient binding of the antigen–antibody complex to the protein A Sepharose beads. The efficiency of antigen–antibody binding can be tested easily by comparing the amount of the antigen present in the cell extract before and after the addition of the primary antibody.

Protocol 7. Testing immunoprecipitation efficiency

1. Establish and maintain cultures as described in *Protocol 3*.
2. Extract the cells with an appropriate extraction buffer. Fractionate soluble from the insoluble residue and process the pellet as described in *Protocol 5*.
3. Prior to immunoprecipitation remove an aliquot (50 μl) of each fraction and quick-freeze.
4. Proceed to immunoprecipitation (*Protocol 6*) through step 8.
5. Before aspirating remove 50 μl of the supernatant from the first wash. Quick-freeze the aliquot if necessary.
6. Quickly thaw the samples, add 15 μl of 4 × SDS sample buffer to each aliquot, vortex, and heat in boiling water for 5 min.
7. Process the samples for SDS-PAGE and Western blotting (50).
8. Compare the relative amount of the protein present in the cell extract before and after the addition of the antibody. If all the antigen present in the cell extract is precipitated with the antibody, there should be very little protein present in the aliquot removed from the cell extract at step 5 of the above protocol. The presence of a significant amount of the protein in this aliquot indicates either inefficient antigen–antibody binding or inefficient binding of the protein A–Sepharose beads to the antigen–antibody complex. The latter however, can be easily determined by *Protocol 8*.

Protocol 8. Determining the efficiency of binding between antigen–antibody complex and protein A–Sepharose beads

1. Establish and maintain cultures as described in *Protocol 3*.
2. Pulse-label and extract as described in *Protocols 4* and *5*.
3. Proceed to immunoprecipitation (*Protocol 6*) and follow through step 8.
4. Do not aspirate supernatant from the first wash. Carefully remove supernatant without disturbing the beads. Transfer to a new tube.
5. Process the beads for washes as described in *Protocol 6*.
6. Add 25–35 μl fresh protein A beads to the supernatant and follow the immunoprecipitation procedures (*Protocol 6*) through step 8.

7. Repeat steps 4–6 of this protocol.
8. Resuspend the final pellets in 45 μl 1 × SDS sample buffer and process for SDS-PAGE and fluorography (*Protocol 9*).
9. Compare the amount of the radioactively labelled protein of interest in each successive immunoprecipitation. If binding of the protein A–Sepharose beads to the antigen–antibody complex is very efficient, more than 95% of the protein is present in the cell extract from the first immunoprecipitation and very little or none in the two successive immunoprecipitations. If significant amount of the relative radioactivity is present in the samples following the first immunoprecipitation, increase the amount of the protein A-Sepharose beads in the step 3 of the above protocol and repeat steps 4–9. If efficiency appears unchanged it could indicate the presence of unbound antigen (see *Protocol 7*). In this case increase the amount of primary antibody used in the initial step.

Protocol 9. SDS-PAGE and fluorography

1. Perform SDS-PAGE as described by Laemmli (34) or by a gel system suitable for the protein of interest.
2. Upon completion of the run stain the gel in Coomassie blue for at least 1 h followed by 2 h in destaining solution.
3. Discard destaining solution and incubate gel in 100% DMSO for 1 h. DMSO can be reused up to three times.
4. Remove DMSO and incubate gel in 20% PPO (Sigma) in DMSO for at least 2.5–3 h. Be careful not to get any moisture in the dish, since this will result in precipitation of PPO in the gel. For high concentrations of acrylamide, PPO/DMSO incubation can go for several hours (i.e. overnight), but for gels with lower concentration (e.g. 5–7%) do not incubate more than 5 h. PPO/DMSO solution can be reused up to three times. DMSO and PPO/DMSO wastes are toxic and should be discarded appropriately. Alternatives to PPO are commercially available (e.g. Enhance, New England Nuclear). They are simpler to use; however, they cost much more than PPO and in our hands haven't worked as well.
5. Remove gels from the PPO/DMSO solution and wash gently with water for 45 min.
6. Dry the gel on to filter paper.
7. If molecular weight markers are run along the samples, mark their positions with radioactive ink (to make labelled ink put a small amount of india ink in a [^{35}S]methionine vial after it is used) on the filter paper along the gel.
8. Expose to autoradiographic films (e.g. Kodak XAR-5).

Biochemical approaches for analysing de novo assembly

The relative amount of radioactivity in each band can be quantified by scanning densitometry or counting the amount of radioactivity (35) present in each band. If scanning, scan several different exposures to obtain accurate measurement of the activity.

Protocol 10. Counting radioactivity from excised gel bands

1. Using an autoradiograph as a template, excise the region of the gel that contains the radioactive protein of interest.
2. Place the gel in a counting vial and rehydrate the gel slice in a small amount of distilled water (20 μl per 1 × 5 mm slice) for 30 min at RT.
3. Incubate the gel slice in 30% H_2O_2/14.8 M ammonium hydroxide in a (19:1) ratio overnight at 37°C (36). Add aqueous-based scintillant. Alternatively, prepare fresh ammonia/solubilizer/scintillant by adding 0.5 ml of 14.8 M ammonium hydroxide to 9.5 ml of NCS solubilizer (Amersham) or Solvable (New England Nuclear), mixing, and adding to 100 ml of toluene-based scintillant. Incubate the gel slice in 10 ml of the above solution overnight at room temperature with shaking (37).
4. Count in scintillation counter.

3.2 Solubility properties of protein components of the junctional complex

The solubility of a protein is defined by its extractability in a specific buffer. In the case of the desmosomal components, we have defined the solubility by their extractability in CSK buffer, a buffer containing Triton X-100 and a high-salt concentration. By definition, the CSK soluble extract includes all the proteins that are not associated with cytoskeleton. Because desmosomal components are linked to the intermediate filament system, one can think of CSK soluble desmosomal proteins as unassembled proteins. Upon synthesis, all of the desmosomal components (DPI/II, DGI, DGII/III, and DPIII) enter a pool of protein which is soluble in CSK buffer. The proteins in the soluble pool subsequently become insoluble. The kinetics of this transfer between the soluble and insoluble pools, however, depends on the culture conditions, i.e. presence or absence of cell–cell contact and each individual protein. Using specific extraction buffer following metabolic labelling, one can analyse the fate of the newly synthesized proteins in the soluble and insoluble pools under different conditions.

3.3 Fate of the newly synthesized proteins following metabolic labelling

Turnover rate and stability of newly synthesized proteins are determined under two conditions: the absence of cell–cell contact and junctional complex assembly, and the presence of cell–cell contact which induces the assembly of junctional complexes (see *Figure 3* and refs 24, 27, 38, 39).

Protocol 11. Determining the fate of the newly synthesized proteins in the absence of cell–cell contact

1. Establish and maintain cultures in 35-mm dishes in the absence of cell–cell contact for at least 3 days as described in *Protocol 1*.
2. Remove the LCM and rinse dishes twice with 1 ml of LCM-MET or HDF or PBS.
3. Incubate each dish in 2.5 ml LCM-MET for 30 min at 37°C.
4. Label each dish with 500 µl of LCM-MET containing the appropriate amount of [^{35}S]methionine (100–150 µCi). Incubate for 15 min at 37°C.
5. Remove the labelling media and rinse dishes three times with HDF.
6. Incubate all but one dish in the LCM-chase at 37°C for desired length of time. For turnover studies six time-points are sufficient, including the initial 0 min chase. The length of the chase period depends on the protein under study. To begin with, 1 h, 3 h, 6 h, 9 h, and 12 h time-points are appropriate. If a protein appears stable, longer chase is required and time-points should be spaced accordingly, such as 0 min, 2 h, 6 h, 10 h, 14 h, and 24 h. To determine the exact time of post-translational modifications such as glycosylation, phosphorylation, changes in solubility (see Section 3.4), a short pulse-chase in conjunction with proper inhibitors is used. In this case, replicate cultures are metabolically labelled and chased in the presence or absence of inhibitors for different lengths of time up to 4 h (e.g., 0 min, 15 min, 30 min, 1 h, 2 h, and 4 h).
7. At the end of each time-point remove chase medium and proceed to extraction as described in *Protocol 5*.

Protocol 12. Determining the fate of the newly synthesized proteins upon induction of cell–cell contact

1. Establish and maintain the cultures as described in *Protocols 1 and 3*.
2. Proceed as described in *Protocol 11* up through step 5.
3. Following the HDF rinse, incubate all but one dish in HCM-chase at 37°C for desired length of time.
4. At the end of each time-point remove the chase medium and proceed to extraction, immunoprecipitation, SDS-PAGE, and fluorography (*Protocols 5–9*).

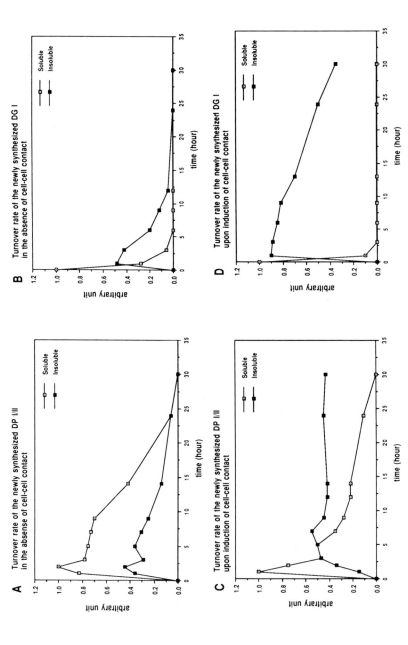

Figure 3. Rate of turnover of DPI/II (**A, C**) and DGI (**B, D**) in confluent monolayers of MDCK epithelial cells in the absence (**A, B**) and presence (**C, D**) of cell–cell contact. Confluent cultures of MDCK epithelial cells were established in LCM and pulsed with 125 μCi [^{35}S]methionine for 15 min (see *Protocol 4*). Duplicate cultures were chased in either LCM (**A, B**) or HCM (**C, D**) for up to 30 h. Soluble (open squares) and insoluble (solid squares) fractions were obtained by extraction in CSK buffer and processed for immunoprecipitation with either DPI/II (**A, C**) or DGI (**B, D**) antibodies. The rate of decrease in levels of [^{35}S]methionine-labelled DPI/II and DGI was determined by fluorography and scanning densitometry.

3.4 Intracellular processing and transport of newly synthesized membrane core glycoproteins

The biosynthetic pathway of a membrane glycoprotein can be analysed by using inhibitors of glycoprotein processing and transport in conjunction with metabolic labelling and pulse-chase experiments (*Figure 4*, and ref. 27). The optimum concentration and incubation time for inhibitors must be determined

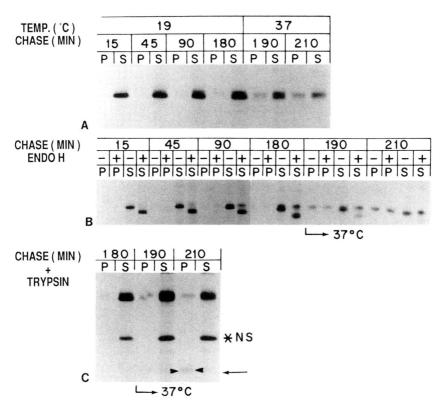

Figure 4. Synthesis, processing and transport of DGI in MDCK epithelial cells at 19°C. Confluent monolayers of MDCK cells were established and maintained in LCM for 16 h. (**A**) Cells were pulse-labelled with 125 μCi [^{35}S]methionine in LCM at 37°C for 10 min and were transferred immediately to 19°C for a chase period of up to 180 min. Duplicate plates were then returned to 37°C and the chase was continued for another 10 or 30 min. Upon completion of the chase period, cells were extracted with CSK buffer and the resulting soluble (S) and insoluble (P) fractions were processed for immunoprecipitation with DGI antisera followed by SDS 5% PAGE and fluorography. (**B**) In duplicate cultures immuno-precipitated samples were divided in half, and one-half was digested with endo-H(+) as described in the text and the other half was used as control (−). (**C**) Duplicate cultures from the last three time-points (180, 190, and 210 min) were treated with trypsin and then processed for immunoprecipitation with DGI antisera and fluorography. *Arrow* indicates the appearance of the 100 kd tryptic fragment of DGI. NS = non-specific protein co-immunoprecipitated with DGI antisera.

Biochemical approaches for analysing de novo assembly

for the cell type and culture conditions used. In the following section these parameters are specifically determined for MDCK cells.

Tunicamycin inhibits the core glycosylation by blocking the synthesis of the dolichol pyrophosphate-linked oligosaccharide precursor that is required for core glycosylation of proteins in the endoplasmic reticulum (40–42). Core glycosylated proteins can be easily detected by their sensitivity to endoglycosidase H (endo-H) digestion. Endo H is an enzyme which hydrolyses high mannose carbohydrate residues. Glycoproteins become resistant to endo-H digestion after trimming of the high mannose precursor oligosaccharides and addition of the first terminal GlcNac residue which indicates complex glycosylation (43).

Protocol 13. Inhibition of core glycosylation

1. Establish and maintain confluent cultures in the absence of cell–cell contact as described in *Protocols 1* and *3*.
2. Prepare stock solutions of tunicamycin in DMSO, and store at $-20\,°C$.
3. Pre-incubate duplicate cultures in LCM containing 33 µM tunicamycin for 2 h.
4. In the last 30 min of incubation replace LCM + tunicamycin with LCM-MET + tunicamycin.
5. Pulse-label and chase in media containing the same concentration of tunicamycin (*Protocol 4*).
6. Following the chase period extract the cells and process for immunoprecipitation (*Protocols 5* and *6*).
7. For endo-H digestion, follow the immunoprecipitation protocol (*Protocol 6*) through washes.
8. Resuspend the final pellet of beads containing the antigen–antibody complex in 200 µl of endo-H incubation buffer and vortex.
9. Divide the samples in half. To one half add 2.5–5 milli-unit endo-H (Boehringer Mannheim); the other half will be control samples without endo-H.
10. Incubate the samples overnight (18 h) at 4°C.
11. Add 900 µl of low-salt immunoprecipitation wash buffer to each sample.
12. Vortex and spin at 4°C in microfuge (14 000 g) for 2 min.
13. Wash twice with low-salt immunoprecipitation wash buffer.
14. Resuspend final pellet in 1 × SDS sample buffer and heat in boiling water for 5 min.
15. Load on to gels and process for fluorography.

Comparison of the relative mobility of the protein immunoprecipitated from control and endo-H-treated cultures in the presence or absence of tunicamycin reveals the time and the size of the glycosylation of the newly synthesized glycoprotein. At early time-points (<45 min), before the newly synthesized core glycosylated protein is transported to the Golgi complex, it is endo-H-sensitive as indicated by the appearance of a lower molecular mass band in endo-H-treated cultures. Once the glycoprotein passes through the Golgi, it becomes complex glycosylated (see Section 3.4) and resistant to endo-H digestion. The appearance of a higher molecular mass endo-H-resistant band at a specific time-point indicates the time of complex glycosylation, and the difference in the molecular masses of the endo-H-sensitive and -resistant proteins determines the size of the glycosylation. In the presence of tunicamycin, however, the control and endo-H-treated cultures do not show a difference in the relative mobility of the immunoprecipitated protein due to the inhibition of core glycosylation by tunicamycin.

To analyse the processing and transport of membrane glycoproteins in the Golgi complex, a reduced temperature (19°C) assay is used in conjunction with metabolic labelling and pulse-chase analysis (27, 44, 45). Incubation of cultures at 19°C immediately following pulse-labelling slows down the transport of glycoproteins between stacks of the Golgi complex and the cell surface, allowing a more detailed analysis of events that occur in the Golgi complex (see *Figure 4* and ref. 27).

Protocol 14. Processing and transport in the Golgi complex: the 19° block

1. To provide 19°C environment, place a regular incubator in a cold room and adjust the temperature to 19°C at least 24 h prior to experiment.

2. Establish and maintain confluent cultures in the absence of cell–cell contact as described in *Protocols 1* and *3*.

3. Pulse-label for 5–10 min.

4. Chase in the appropriate medium, and incubate duplicate cultures at 37°C or 19°C. Pre-cool the 19°C chase medium.

5. Upon completion of the chase period, process cultures for immunoprecipitation, SDS-PAGE, and fluorography (*Protocols 6* and *9*).

6. For detection of complex glycosylation, use endo-H in duplicate samples as described in *Protocol 13*.

For studies examining vesicular transport between the Golgi complex and plasma membrane, inhibitors such as monensin are used in metabolic labelling and pulse-chase experiments (46–48).

Protocol 15. Transport between Golgi complex and plasma membrane

1. Establish and maintain confluent cultures in the absence of cell–cell contact as described in *Protocols 1* and *3*.
2. Prepare stock solutions of monensin (Sigma) in DMSO and store at $-20\,°C$.
3. Incubate cultures in LCM-MET containing 25 µM monensin for 30 min.
4. Pulse-label and chase in the appropriate media containing the same concentration of monensin.
5. Trypsinize or biotinylate (see *Protocol 17*) duplicate cultures at specific chase times for detecting the arrival of the glycoprotein at the plasma membrane.
6. Upon completion of the chase period extract the cultures and process for immunoprecipitation, SDS-PAGE, and fluorography (*Protocols 6* and *9*). Check the autoradiograms for the presence of the tryptic fragment or biotinylated protein in control and monensin-treated cultures. In monensin-treated cultures the majority of the newly synthesized glycoprotein is trypsin-resistant and remains intracellular. Similarly, in these cultures the newly synthesized glycoprotein can not be biotinylated.

Several techniques can be used to detect the presence of glycoproteins at the cell surface. The following describes two simple procedures, trypsinization (27) and biotinylation (49, 50) of cell surface proteins, used in conjunction with metabolic labelling and pulse-chase studies (see *Figure 5*).

Protocol 16. Detection of glycoproteins at the plasma membrane: extracellular trypsinization

1. Establish and maintain confluent cultures in the absence of cell–cell contact (*Protocols 1* and *3*).
2. Pulse-label and chase duplicate cultures (control and trypsin-treated) for each time-point in appropriate media.
3. At specific time-points during the chase period remove media and rinse cultures three times with PBS or HDF. Process the control cultures as described for immunoprecipitation.
4. For trypsin treatment, add 1 ml of 0.125% trypsin (Difco 1:250 or TPCK-treated) in HDF and incubate at 37°C. The incubation time varies with culture conditions. Cultures must be monitored by light microscopy to determine the optimum incubation time. Stop incubation when cells start to round up.
5. Inactivate trypsin by adding FBS (250 µl), soybean trypsin inhibitor (4 mg/ml), or PMSF (2 mM).

6. Save the trypsin supernatant if required. This supernatant contains the extracellular domain of the glycoprotein. The cytoplasmic domain is protected by the plasma membrane. Process the supernatant for immunoprecipitation (*Protocol 6*).
7. Rinse the cultures with HDF and proceed for extraction and immunoprecipitation (*Protocols 5* and *6*).
8. Load the samples, control next to trypsin-treated for each time-point, on to appropriate gels and process for SDS-PAGE and fluorography (*Protocol 9*).
9. Compare the size of the protein immunoprecipitated from the control and trypsin-treated cultures for each time-point. The appearance of the lower-molecular-weight tryptic fragment at a specific time-point indicates the time of arrival of the glycoprotein at the cell surface.

Protocol 17. Detection of glycoproteins at the plasma membrane: cell surface biotinylation

1. Establish and maintain cultures at confluent density in the absence of cell–cell contact (*Protocols 1* and *3*).
2. Pulse-label and chase as required.
3. Following the chase period rinse the cultures three times with HDF.
4. Incubate the cultures in 5 mM EGTA in HDF or PBS for 10 min at RT on a rocker platform.
5. Rinse five times with Hepes–saline buffer.
6. Incubate each 35 mm Petri dish with 1.5 ml of 300 µg/ml NHS-SS-Biotin (Pierce) in Hepes–saline buffer for 15 min at RT on a rocking platform.
7. Wash three times with Tris–saline/PMSF.
8. Extract and process for immunoprecipitation. Follow the immunoprecipitation procedure in *Protocol 6* with the following exceptions:
9. Wash the beads containing the antigen antibody complex with HS ± sucrose, high-salt, and low-salt buffers *without DTT*. Thiols such as DTT would result in the cleavage of the disulfide bond and release the antigen–antibody complex from the biotin complex.
10. Resuspend the final pellet in 50 µl of 2% SDS and heat in boiling water for 3 min.
11. Spin and decant the supernatant into clean tubes. Discard the beads.
12. Dilute the supernatant to 0.1% SDS with HS-buffer *without DTT*.
13. Add 45 µl of avidin–agarose beads (Pierce) and incubate on a rocker rotator at 4°C for 1.5 h.

Biochemical approaches for analysing de novo assembly

Protocol 17. *Continued*

14. Wash 3× with immunoprecipitation wash buffers *without DTT* as described above.
15. Resuspend the final pellet in 1 × SDS sample buffer *containing 100 mM DTT* and heat in boiling water for 3 min.
16. Process samples for SDS-PAGE and fluorography (*Protocol 9*).

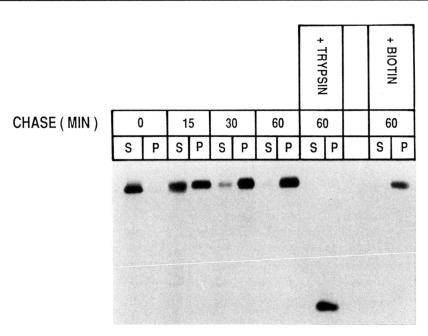

Figure 5. Detection of the newly synthesized DGI at the plasma membrane of MDCK cells upon induction of cell–cell contact. Confluent monolayers of MDCK cells were pulse labelled with 125 µCi [^{35}S]methionine in LCM and chased for 0, 15, 30, and 60 min. Duplicate cultures for 60-min time-point were either trypsinized or biotinylated as described in the text. Cells were extracted with CSK buffer and the soluble (S) and insoluble (P) fractions were processed for immunoprecipitation with DGI antisera followed by SDS 5% PAGE and fluorography.

4. Conclusion

This chapter has described the basic methodology for analysing *de novo* assembly of junctional components by studying the kinetics of synthesis, processing, transport, and stability of constituent proteins. Establishment of 'contact naïve' cultures prior to assembly studies is of critical importance and can be achieved by modulating the Ca^{2+} concentration of the growth media as described in *Protocol 1*. It must be noted, however, that the Ca^{2+} concentrations given in this chapter are determined for MDCK epithelial cells specifically, and may not be appropriate for other epithelial cell lines (for example, the optimum low Ca^{2+} concentration media for keratinocytes is 50 μM). Metabolic labelling studies described here are basic [^{35}S]methionine labelling procedures which can be modified according to the radioactive precursors used and the type of studies required. For example, for phosphorylation studies where cells are metabolically labelled with [^{32}P]orthophosphate the preincubation media should be phosphate-free and the labelling time considerably longer (see ref. 51). When using inhibitors, the optimum effective concentration and incubation time must be determined. Appropriate controls must be done for each step and the specificity of the antibodies as well as efficiency of immunoprecipitation must be determined, particularly for comparative kinetic studies of proteins.

Acknowledgements

I would like to thank Drs Ann Acheson and Teresa Krukoff for their critical reading of this manuscript, and Dr W. James Nelson for his contributions towards the information contained in this chapter. Many thanks to Lynn Frasch for her patience and excellent typing of this manuscript and all its revisions.

Appendix: Constitution of media, buffers and solutions

All chemicals are purchased from Sigma unless stated otherwise

Media

Low [Ca^{2+}] DMEM—methionine, serum, and antibiotics free

Components	mg/litre
KCl	400.00
$MgSO_4.7H_2O$	199.80
NaCl	5962.80
D-Glucose (dextrose, mono)	1000.00
Phenol red	10.00

Amino Acids	
L-Arginine-HCl	126.00
L-Cystine.2HCl	31.23
L-Glutamine	295.87
L-Histidine HCl.$2H_2O$	42.00
L-Isoleucine	52.00
L-Leucine	52.00
L-Lysine HCl	72.50

L-Phenylalanine	32.00
L-Threonine	48.00
L-Tryptophan	10.00
L-Tyrosine	51.98
L-Valine	45.83

[Amino acids need to stir at least 1 h (1–24) to dissolve.]

100 × vitamin solution (Gibco) 10 ml	
$NaHCO_3$	1000.00
Na Hepes	2603.00
$NaH_2PO_4 \cdot H_2O$	140.00
or	
NaH_2PO_4 (anhydrous)	120.00
$CaCl_2 \cdot 4H_2O$	0.742

Adjust the pH to 7.0 and filter sterilize with 0.22 μM filter.

100 × Methionine solution

L-Methionine	0.15 g
Distilled water	100 ml

Filter sterilize with 0.22 μM filter

100 × Antibiotics solution
Dissolve in 100 ml of PBS:

Kanamycin sulfate	1.22 g
Penicillin 'G' sodium	0.3 g
(100 million units)	
Streptomycin sulfate	0.50 g

Filter sterilize with 0.22 μM filter

LCM

Low [Ca^{2+}] DMEM	100 ml
Methionine	1.5 mg
Antibiotics (100×)	1 ml
DFBS*	10 ml

(*Dialyse 100 ml of FBS against 4 litres of PBS or Tris–saline with two changes each day for 2 days at 4°C.)

LCM-MET

Low [Ca^{2+}] DMEM	100 ml
Antibiotics (100×)	1 ml
DFBS	2.5 ml

High Ca^{2+} Concentration DMEM

1 package (1 L) of DMEM (Eagle's serum-free, Gibco)	
sodium bicarbonate	1.0 g
1 litre of distilled water	

Adjust the pH to 7.0 and sterilize
add:
10% FBS
1% (100×) antibiotics

HCM-chase
HCM	100 ml
Methionine (100×)	1 ml

LCM-chase
LCM	100 ml
Methionine (100×)	1 ml

Tris–saline
15 mM Tris–HCl pH 7.5
120 mM NaCl

Tris–saline/PMSF
Tris–saline + 0.1 mM PMSF

HDF
137 mM	NaCl
5.4 mM	KCl
0.1%	Glucose (dextrose monohydrate)
4 mM	NaHCO$_3$
0.5 mM	EDTA

Filter sterilize

PBS
20 mM	sodium phosphate
120 mM	NaCl

Filter sterilize

Dulbecco's PBS
0.9 mM	CaCl$_2$
2.7 mM	KCl
1.5 mM	KH$_2$PO$_4$
0.5 mM	MgCl$_2$
137 mM	NaCl
8.1 mM	Na$_2$HPO$_4$

Filter sterilize

CSK (cytoskeleton) extraction buffer
300 mM	Sucrose
50 mM	NaCl
10 mM	Pipes, pH 6.8
3 mM	MgCl$_2$
0.5% (v/v)	Triton X-100

1.2 mM	PMSF
0.1 mg/ml	DNase (Boehringer Mannheim)
0.1 mg/ml	RNase (Boehringer Mannheim)

[$(NH_4)_2SO_4$ to a final concentration of 250 mM; added 10 min into extraction, see *Protocol 5*]

High-stringency wash buffer (HS-buffer)
0.1%	SDS
1%	Na deoxycholate
0.5%	Triton X-100
20 mM	Tris–HCl pH 7.5
120 mM	NaCl
25 mM	KCl
5 mM	EDTA
5 mM	EGTA
0.1 mM	DTT

Filter through 0.45 μM filter

HS-buffer + sucrose
HS-buffer
1 M Sucrose

High-salt immunoprecipitation buffer
HS-buffer
1 M NaCl

Low-salt immunoprecipitation buffer
10 mM	Tris–HCl pH 7.5
2 mM	EDTA
0.5 mM	DTT

Filter through 0.45 μM filter

Endo-H incubation buffer
20 mM	sodium phosphate pH 6.0
20 mM	NaCl

Hepes–saline buffer (for biotinylation)
10 mM	Hepes pH 7.4
154 mM	NaCl
7.2 mM	KCl

Staining solution
0.1%	Coomassie brilliant blue
10%	glacial acetic acid
50%	ethanol
40%	distilled water

Dissolve Coomassie brilliant blue in ethanol and filter prior to mixing

Destaining solution
12.5%   ethanol
5%   glacial acetic acid
82.5%   distilled water

References

1. Farquhar, M. G. and Palade, G. E. (1963). *J. Cell Biol.*, **17**, 375.
2. Gumbiner, B., Lowenkopf, T., and Apatira, D. (1991). *Proc. Natl Acad. Sci. USA*, **88**, 3460.
3. Stevenson, B. R. and Goodenough, D. A. (1984). *J. Cell Biol.*, **98**, 1209.
4. Geiger, B., Avnur, Z., Volberg, T., and Volk, T. (1985). In *The Cell in Contact*, (eds. G. M. Edelman and J. P. Thiery) pp. 461–89. John Wiley and Sons, Inc. New York.
5. Volk, T. and Geiger, B. (1984). *EMBO J.*, **3**(10), 2249.
6. Garrod, D. R. and Cowin, P. (1986). In *Receptors in Tumor Biology* (ed. C. M. Chadwicic), pp. 95–130. Cambridge University Press, Cambridge.
7. Kelly, D. E. (1966). *J. Cell Biol.*, **28**, 51.
8. Overton, J. (1974). *Prog. Surf. Membr. Sci.*, **8**, 161.
9. Gallin, W. J., Sorkin, B. C., Edelman, G. M., and Cunningham, B. A. (1987). *Proc. Natl Acad. Sci. USA*, **84**, 2808.
10. Takeichi, M. (1990). *Annu. Rev. Biochem.*, **59**, 237.
11. Cohen, S. M., Gorbsky, G., and Steinberg, M. S. (1983). *J. Biol. Chem.*, **258**(4), 2621.
12. Cowin, P., Kapprell, H. P., Franke, W. W., Tamkun, J., and Hynes, R. O. (1986). *Cell*, **46**, 1063.
13. Duden, R. and Franke, W. W. (1988). *J. Cell Biol.*, **107**, 1049.
14. Franke, W. W., Schmid, E., Grund, C., Muller, H., Engelbrecht, H., Moll, R., Stadler, J., and Jarasch, E. D. (1981). *Differentiation*, **20**, 217.
15. Gorbsky, G. and Steinberg, M. S. (1981). *J. Cell Biol.*, **90**, 243.
16. Skerrow, C. J. (1985). In *Biology of the Integument*, Vol. 2, *Vertebrates*, (eds. J. Bereiter-Hahn, A. G. Matoltsy, and K. S. Richards) pp. 763–87. Springer-Verlag, Berlin.
17. Skerrow, C. J., Hunter, I., and Skerrow, D. (1987). *J. Cell Sci.*, **87**, 411.
18. Skerrow, C. J. and Matoltsy, A. G. (1974). *J. Cell Biol.*, **63**, 515.
19. Jones, J. C. R. and Goldman, R. D. (1985). *J. Cell Biol.*, **101**, 506.
20. Hennings, H. and Holbrook, K. A. (1983). *Exp. Cell. Res.*, **143**, 127.
21. Hennings, H., Michael, D., Cheng, C., Steinert, P., Holbrook, K., and Yuspa, S. H. (1980). *Cell*, **19**, 245.
22. Mattey, D. L. and Garrod, D. R. (1986). *J. Cell Sci.*, **85**, 95.
23. Watt, F. M., Mattey, D. L., and Garrod, D. R. (1984). *J. Cell Biol.*, **99**, 2211.
24. Pasdar, M. and Nelson, W. J. (1988). *J. Cell Biol.*, **106**, 677.
25. Madin, S. J. and Darby, N. B. (1979). *American Type Culture Collection Catalogue of Strains* II, p. 30.
26. Pasdar, M. and Nelson, W. J. (1988). *J. Cell Biol.*, **106**, 687.
27. Pasdar, M. and Nelson, W. J. (1989). *J. Cell Biol.*, **109**, 163.
28. Pasdar, M., Krzeminski, K. A., and Nelson, W. J. (1991). *J. Cell Biol.*, **113**, 645.

29. Nelson, W. J. and Veshnock, P. J. (1986). *J. Cell Biol.*, **103**, 1751.
30. Bornstein, M. B. (1958). *Lab. Invest.*, **7**, 134.
31. Ehrmann, R. L. and Gay, G. O. (1956). *Natl Cancer Inst. J.*, **16**, 1375.
32. Fey, E. G., Wan, K. M., and Penman, S. (1984). *J. Cell Biol.*, **98**, 1973.
33. Towbin, H., Staehlin, T., and Gordon, J. (1979). *Proc. Natl Acad. Sci. USA*, **76**, 4350.
34. Laemlli, U. K. (1970). *Nature, London*, **227**, 680.
35. Laskey, R. A. and Mills, D. (1975). *Eur. J. Biochem.*, **56**, 335.
36. Bonner, W. M. and Laskey, R. A. (1974). *Eur. J. Biochem.*, **46**, 83.
37. Ward, S., Wilson, D. L., and Gilliam, J. J. (1970). *Anal. Biochem.*, **38**, 90.
38. Penn, E. J., Burdett, I. D. J., Hobson, C., and Magee, A. I. (1987). *J. Cell Biol.*, **105**, 2327.
39. Penn, E. J., Hobson, C., Rees, D. A., and Magee, A. I. (1987). *J. Cell Biol.*, **105**, 57.
40. Czichi, U. and Lennarz, W. J. (1977). *J. Biol. Chem.*, **252**, 7901.
41. Hubbard, S. C. and Ivatt, R. J. (1981). *Annu. Rev. Biol. Chem.*, **50**, 555.
42. Tkacz, J. S. and Lampen, J. O. (1975). *Biochem. Biophys. Res. Commun.*, **65**, 248.
43. Tarentino, A. L., Trimble, R. B., and Maley, F. (1978). *Methods in Enzymology*, Vol. 50 (ed. V. Ginsburg), pp. 574–80. Academic Press, Orlando, Florida.
44. Dunphy, W. G., Brands, R., and Rothman, J. E. (1985). *Cell*, **40**, 463.
45. Matlin, K. S. and Simons, K. (1983). *Cell*, **34**, 283.
46. Johnson, D. C. and Spear, P. G. (1982). *J. Virol.*, **43**(3), 1102.
47. Tartakoff, A. M. (1983). *Cell*, **32**, 1026.
48. Tartakoff, A. M. and Vassalli, P. (1977). *J. Exp. Med.*, **146**, 1332.
49. Lisante, M. D., Sargiacomo, M., Graeve, L., Saltiel, A. R., and Rodriguez-Boulan, E. (1988). *Proc. Natl Acad. Sci. USA*, **85**, 9557.
50. Matter, K., Brauchbar, K., and Hauri, H. P. (1990). *Cell*, **60**, 429.
51. Parrish, E. P., Maston, J. E., Mattey, D. L., Measures, H. R., Venning, R., and Garrod, D. R. (1990). *J. Cell Sci.*, **96**, 239.

10

Biochemical methods for studying supramolecular complexes involving cell adhesion molecules, integral membrane proteins, and the cytoskeleton

W. JAMES NELSON, RACHEL WILSON, and ROBERT W. MAYS

1. Introduction

During development of complex tissue patterns, cells undergo dramatic structural and functional reorganization (1). The inductive process(es) leading to cellular differentiation often involves specific cell–cell and cell–substratum interactions (2, 3). These interactions result in new programmes of gene expression and the spatial redistribution of plasma membrane and cytoplasmic proteins in line with the specialized structural and functional requirements of the fully differentiated cell. Understanding the mechanisms involved in the regulation of protein distributions during cellular differentiation is one of the most interesting and experimentally challenging areas of cell and developmental biology.

Epithelial cell differentiation provides an important and useful system to analyse mechanisms involved in the remodelling of protein distributions following induction of cell–cell and cell–substratum contacts. During development of kidney epithelium, expression of specific cell adhesion molecules (CAMs; reviewed in ref. 4), in particular E-cadherin, induces the aggregation (condensation) of precursor mesenchymal cells into cell clusters (2). Together with inductive signals through cell–substratum interactions (laminin A; ref. 4), cell–cell adhesion results in the gradual development of epithelial cell polarity that is reflected in the reorganization of the plasma membrane into three structurally and functionally distinct cell surface domains, termed apical, lateral and basal, although the latter two domains are often collectively termed basal–lateral (reviewed in refs 1 and 5). Each domain has a characteristic composition of lipids and integral membrane and cytoplasmic proteins, some of which were constitutively expressed in the non-polarized precursor cells and others were the products of an epithelial-specific programme of gene expression (1, 5). These changes in cell organization induced by cell–cell and cell–substratum adhesion can also be followed *in vitro* using several established cell lines derived from kidney and intestinal epithelium (1, 5).

How do we approach experimentally the mechanisms involved in the structural and functional reorganization of these cells as a consequence of cell–cell adhesion? Other chapters in this book address how mechanisms involved in cell–cell adhesion can be approached. This chapter will focus on methods for analysing the consequences of cell–cell adhesion on protein distributions in cells. An integrated approach is described in which changes in protein distributions are monitored by immunofluorescence microscopy, and protein interactions are analysed by subcellular fractionation and the isolation of protein complexes. This approach was developed to analyse interactions between membrane proteins and elements of the cytoskeleton during the development of epithelial cell polarity using an *in vitro* model of epithelial cells, Madin–Darby canine kidney (MDCK) cells (6–9). We hope that they are also adaptable for the investigation of different protein complexes in other cell types.

1.1 General considerations

Prior to analysis of protein distributions and complexes during cell–cell adhesion in epithelial cells, it is useful to consider the following points that will help in determining the experimental approach that will be followed:

- *Homogeneity of cell population*. The homogeneity of the cells used for analysis is not critical, but can greatly simplify the interpretation of results.

- *Synchrony of change(s) in cell organization*. The degree of synchrony in the induction of changes in cellular organization being investigated will affect the simplicity of the interpretation of results.

- *Quantity of cells that can be provided for analysis*. The quantity of cells available will affect the scope of the analysis; for instance, the production of antibodies from unknown components of an isolated complex will depend upon the amount of protein available which, in turn, will depend upon the quantity of cells used.

- *Means for identification of proteins*. The availability of antibodies to at least one component of the protein complex under investigation will be useful.

- *Subcellular distribution of one or more of the proteins of interest in cells prior to extraction and analysis of complexes*. The identification of proteins either associated with cell–cell contacts, or whose distribution changes upon cell–cell contact provides a useful starting point in the search for protein complexes.

- *Co-extraction of proteins of interest from whole cells.* An important criterion for the analysis of protein complexes is that proteins can be co-extracted from the cells for further fractionation.

Both of the last two points are worth considering prior to tooling-up for analysis of protein complexes. However, in many cases antibodies will not (yet) be available for analysing the subcellular distribution and extraction of proteins. In this case, it is useful to retrospectively analyse these parameters following identification of putative protein complexes in whole-cell extracts as a test of their cellular significance in a whole-cell context.

2. Cells

Although any cell type can be used for the analysis of protein complexes, those complexes associated with cell adhesion in epithelial cells can be conveniently analysed, using a variety of epithelial cell lines which have retained their polarized phenotype. Some of the more commonly used cell lines are:

- *Kidney epithelial cells*: MDCK cells, LLC-PK$_1$ (porcine proximal tubule)
- *Intestinal epithelial cells*: Caco 2, HT-29, T-84 (human intestine)

2.1 Cell culture

Maximum advantage of the polarized epithelial cell lines can be taken by growing the cells on permeable filters. While this is not critical for the analysis of protein complexes, growth of cells on filters, in which the growth medium directly bathes the basal–lateral cell surface, provides a more natural environment than a Petri dish. The basal–lateral surface of polarized epithelial cells contains growth factor receptors and some amino-acid transporters (1, 5). When these cells are grown on a solid substratum (e.g. Petri dish), the tight junction inhibits the diffusion of the growth medium to the basal–lateral surface. There is also an experimental advantage to growing cells on filters. Apical and basal–lateral membrane proteins can be selectively labelled on the cell surface, thus providing a simple method for determining the domain from which different membrane proteins in a protein complex were derived (see Section 2.4).

As an example, protocols will be described for growing MDCK cells. However, these methods are applicable to the culture of other polarized epithelial cells.

Protocol 1. Establishing MDCK cell cultures on filters

Materials

- MDCK cells (can be purchased from American Type Culture Collection)[a]

Protocol 1. *Continued*

- Polycarbonate filter inserts (Millipore, Costar) with 0.45 μm pore size, 2.4 cm diameter[b]
- 6-well tissue culture trays for filter inserts (Costar)

Methods

1. Prepare a single cell suspension of MDCK cells at a density of approx. 2×10^6 cells/ml of Dulbecco's modified Eagle's medium (Gibco) supplemented with 10% fetal calf serum (Gibco) [DMEM/FCS].[c]
2. Gently pipette 1.5 ml of the cell suspension into the filter insert that has been placed in the well of a 6-well culture plate. This cell density will establish a confluent monolayer of cells.
3. Pipette 2.5 ml of DMEM/FCS into the outside compartment of the well surrounding the filter insert. The height of the medium on the inside and outside compartments of the filter should be the same so that there is little or no hydrostatic pressure across the monolayer.
4. Place the culture in a 37°C, humidified, 5% CO_2 in air incubator.

[a] It is assumed that the reader has basic knowledge of sterile tissue culture techniques and has available the following apparatus: sterile, positive pressure, or laminar flow hood; 37°C, humidified 5% CO_2 in air incubator; low speed (1500 g), swinging-bucket centrifuge; phase-contrast microscope.

[b] The filter inserts can be purchased according to pore size, diameter, degree of transparency, and filter composition. The authors' preference is for Costar Transwell polycarbonate filter inserts, 0.45 μm pore size, 2.4 cm diameter. These filters are not transparent, but non-microscopical methods are available to determine whether the monolayer of cells growing on the filter have attained a full degree of cell–cell adhesion and confluency (see *Protocol 3*).

[c] It is useful to pre-culture MDCK cells under conditions in which there is little or no cell–cell contact. These 'contact naïve' cells lose, through normal protein turnover, residual structures from previous cell–cell contacts. Thus, induction of contacts induces a *de novo* response in these cells. To pre-culture cells, prepare a single-cell suspension at approx. $1-2 \times 10^6$ cells/ml DMEM/FCS. Add 1 ml of the cell suspension to 15 ml DMEM/FCS and pipette the cell suspension into a 150-cm diameter plastic tissue culture Petri dish (Nunc). 24 h later, trypsinize the cells from the Petri dish, and repeat the procedure. Following a further 24 h in culture, trypsinize the cells and use them for experimentation.

2.2 Induction of cell–cell adhesion in monolayer of epithelial cells

Protocol 1 can be modified to prepare a homogeneous population of epithelial cells in which cell–cell adhesion can be rapidly and synchronously induced across the monolayer, thus allowing convenient replication of cell–cell contact in multiple cultures (see *Figure 1*). The basis for the protocol is the modulation of extracellular Ca^{2+} concentration in the growth medium. In cells such as MDCK cells, the Ca^{2+}-dependent CAM responsible for initial cell adhesion is E-cadherin (10). E-cadherin requires >250 μM $[Ca^{2+}]_e$ for

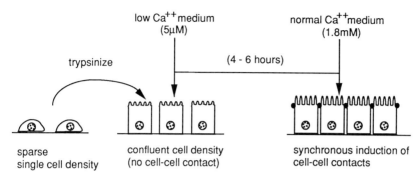

Figure 1. Modulation of cell–cell contact in confluent monolayers of MDCK epithelial cells.

active conformation and cell–cell adhesion (11). When the $[Ca^{2+}]_e$ is below <250 µM, E-cadherin is inactive and there is little or no cell–cell adhesion.

Protocol 2. Establishing synchronous cell–cell adhesion in confluent monolayers of MDCK cells

Materials
As in *Protocol 1* except: DMEM and FCS containing 5 µM Ca^{2+} instead of 1.8 mM Ca^{2+}. DMEM and FCS containing 5 µM Ca^{2+} are not commercially available, but can be prepared in the laboratory from the constituent components of salts, glucose, amino acids, and vitamins according to the constituents for DMEM in the Gibco catalogue, except that 5 µM Ca^{2+} is included instead of 1.8 mM Ca^{2+}. FCS is depleted of Ca^{2+} by extensive dialysis against 4–6 changes of a 100-fold excess of PBS (without Ca^{2+} or Mg^{2+}) for 2–4 days at 4°C; the FCS is then filter sterilized using a disposable 0.2 µm filter apparatus (Corning).

Methods
1. Pre-culture MDCK cells as described in *Protocol 1*, footnote c
2. Following trypsinization of the second round of single-cell density cultures, resuspend the cell pellet in DMEM/FCS containing 5 µM Ca^{2+} to a density of $1.8–2.0 \times 10^6$ cells/ml.
3. Pipette 1.5 ml of the cell suspension into the filter insert, and 2.5 ml DMEM/FBS containing 5 µM Ca^{2+} into the outer compartment as described in *Protocol 1*.
4. Place the cells in an incubator. Approximately 4–7 h are required for cells to attach to the filter substratum.
5. To induce Ca^{2+}-dependent cell–cell adhesion, remove the growth medium by aspiration and replace with normal DMEM/FCS containing 1.8 mM Ca^{2+}

2.3 Monitoring cell–cell adhesion and confluency in monolayers of MDCK cells growing on filter inserts

Cells may be cultured on filters that are transparent, thus allowing direct observation of cells with a phase-contrast microscope. However, for determining whether the monolayer is complete and tight junctions are formed, it is useful to use a quantitative method; this is particularly critical for cell surface labelling of membrane proteins in which either the apical or basal–lateral membrane domain is labelled (see Section 2.4).

Protocol 3. Use of radiolabelled tracers to monitor the structural integrity of cell monolayers growing on filter inserts

Materials
- [^{3}H]-inulin (New England Nuclear; 313 mCi/g)
- basic equipment and apparatus for liquid scintillation spectrometry
- Ringer's buffer solution

Methods
1. Wash filter insert cultures twice in Ringer's buffer solution that has been pre-warmed to 37°C.
2. Add 2 μCi [^{3}H]-inulin to 1.5 ml Ringer's buffer solution, mix well and pipette it into the apical compartment of the filter insert; add 2.5 ml of Ringer's buffer solution to the outside, basal–lateral compartment of the filter insert.[a]
3. Return the culture to the 37°C incubator.
4. At intervals (e.g. 15–30 min), remove 50 μl of the medium from the inside (apical) compartment and outside (basal–lateral) compartment of the filter insert and place in 5-ml scintillation vials.
5. Add 5 ml scintillation fluid.
6. Cap the vials and determine radioactivity by liquid scintillation counting.

[a] Alternatively, [^{3}H]-inulin can be added to the basal–lateral compartment buffers, and diffusion of tracer across the monolayer into the apical compartment determined.

The use of [^{3}H]-inulin for determining the integrity of the monolayer is relatively quick, simple, and sensitive, and does not require specialized equipment beyond that found in most cell biology/biochemistry laboratories. Generally, diffusion is measured over a 30-min period. In cultures in which cell–cell contact was synchronously induced in confluent monolayers by Ca^{2+}-switch (see *Protocol 2*), about 1.5–2.5 h are required for induction of cell–cell contact and tight junction formation such that inulin does not diffuse across the monolayer (see *Figure 2*). Use of the culture should be avoided in

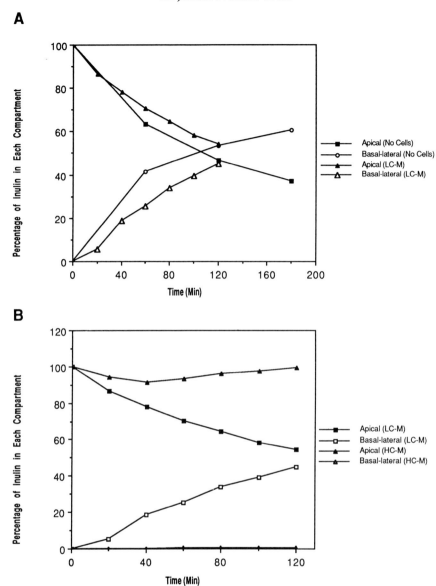

Figure 2. Measurement of cell monolayer and tight junction integrity with [^{3}H]inulin in cultures of MDCK cells following induction of cell–cell contact. [^{3}H]inulin was added either to the apical or basal–lateral compartments for filters. Aliquots of the growth medium were removed from both the apical and basal–lateral compartments and counted. The amount of [^{3}H]inulin in each compartment at a given time is expressed as a percentage of the total amount. (**A**) Comparison of the diffusion of [^{3}H]inulin across the filter in the absence or presence of cells grown in DMEM/FCS containing 5 μm Ca^{2+} (LC-M). (**B**) Comparison of the diffusion of [^{3}H]inulin across the filter containing cells grown in LC-M or normal DMEM/FCS containing 1.8 mM Ca^{2+}.

experiments in which different cell surface domains are to be labelled when more than 0.5% of tracer is found to diffuse across the monolayer.

2.4 Labelling membrane proteins on different cell-surface domains

Growth of MDCK monolayers on filters allows direct access to both the apical and basal–lateral cell-surface domains, and cell confluency and the formation of functional tight junctions result in the physical separation of the apical and basal–lateral compartments. Identification of membrane proteins in protein complexes that were present in either the apical or basal–lateral membrane domain can be determined by cell-surface labelling with biotin (12), which can be added to buffer in either compartment without diffusion across the monolayer (see *Figure 3*). A number of biotinylating agents are available for this type of analysis (*Table 1*).

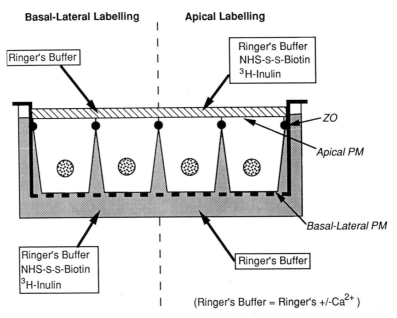

Figure 3. Use of biotinylated reagents to selectively label apical or basal–lateral membrane proteins in confluent monolayers of MDCK cells.

Protocol 4. Cell-surface protein labelling with biotinylating reagents

Materials
- biotinylating reagent (30 mg/ml stock solution); see *Table 1*

- Ringer's buffer solution
- Tris–saline (50 ml Tris–HCl, pH 7.5, 120 mM NaCl)

Methods

1. Test filter-grown, confluent monolayers of MDCK cells for tight junction integrity using [^{3}H]-inulin. Use filter cultures that exhibit <0.5% diffusion of [^{3}H]-inulin across the monolayer.
2. Wash filter cultures twice with ice-cold Ringer's buffer solution, and remove all liquid by aspiration.
3. Dilute freshly prepared stock solution of biotinylated reagent into ice-cold Ringer's buffer solution to give a final concentration of 300 μg/ml.
4. Pipette 1.5 ml or 2.5 ml of Ringer's buffer solution containing biotinylating reagent into *either* the apical or basal–lateral compartment of the filter insert, respectively; add either 2.5 ml or 1.5 ml Ringer's buffer solution to the opposite compartment of the filter insert depending upon which side the biotinylated reagent was added.
5. Place the filters on a rocking platform at 4°C for 30 min.
6. Aspirate solutions from both the apical and basal–lateral compartments of the filter insert.
7. Wash cells three times in ice-cold Tris–saline to block free amine groups on surface-expressed membrane proteins and quench remaining biotinylated reagent.

Cells in which different membrane domains have been cell-surface labelled with biotin can be extracted and used for analysis of components of protein complexes (see below).

3. Preliminary analysis of the cellular organization of proteins prior to complex isolation

An important aspect of the procedures for isolating and characterizing protein complexes is to have an idea of at least some of the proteins that may be present. Such knowledge will become useful in determining whether interactions between proteins in complexes isolated from whole cells are significant in the light of their subcellular localization in intact cells. In addition, since proteins must be extracted from cells for analysis (see Section 3.2), prior knowledge of the solubility of different proteins is particularly useful in determining the extraction protocol. Clearly, if two or more proteins are found in a protein complex, but their subcellular distributions in intact cells are different, the significance of the isolated complex is doubtful.

Table 1. Cell surface biotinylating reagents (available through Pierce Chemicals)

Reagent	Solubility	Remarks
Sulfosuccinimidobiotin [sulfo-NHS-biotin]	water	–
Sulfosuccinimidyl 6-(biotinamido) hexanoate [NHS-LC-biotin]	water	extended spacer
Sulfosuccinimidyl 2-(biotinamido) ethyl-1, 3-dithiopropionate [NHS-SS-biotin]	DMSO	reversible (cleavable spacer arm contains a disulfide bond)

3.1 Subcellular localization of proteins

Indirect immunofluorescence microscopy provides a simple and rapid method to determine the subcellular distributions of proteins in intact cells. In addition, a combination of procedures in which cells are first fixed and then permeabilized, or vice versa, provides a qualitative method for comparing the extractability of different proteins. Proteins are detected with antibodies raised and characterized to specific, purified proteins.

Protocol 5. Indirect immunofluorescence for comparing subcellular distribution and extractability of proteins in intact cells

Materials

- 1.75% (v/v) formaldehyde (37% stock solution) in phosphate-buffered saline (PBS)
- PBS
- Triton X-100 extraction buffer (15 mM Tris–HCl, pH 7.5, 120 mM NaCl, 25 mM KCl, 2 mM EDTA, 2 mM EGTA, 0.1 mM DTT, 0.5 mM PMSF, 0.5% (v/v) Triton X-100)
- blocking buffer (0.2% BSA, 50 mM NH_4Cl, 1% goat serum in PBS) Texas red—or fluorescein (FITC)—conjugated secondary antibodies (affinity purified goat antirabbit or goat anit-mouse)
- Elvanol (20 g Mowiol (Calbiochem) in 80 ml PBS and 40 ml glycerol)
- glass coverslips (heat-sterilized)
- 3.5-cm diameter plastic tissue culture Petri dish (Bellco)
- glass microscope slides

Method A. Cell culture

1. Prepare a single cell suspension of MDCK cells (approx. 1.5×10^6 cells/ml) in DMEM/FCS containing 5 μM Ca^{2+} as described in *Protocol 2*.

Karl J. Karnaky, Jr.

Protocol 5 details a basic procedure for measuring the flux of an isotope across an epithelium mounted in an Ussing chamber in the short-circuit condition. As pointed out earlier (Section 2), the I_{sc} is the algebraic sum of all active net ionic fluxes across the epithelium. For some epithelia (for example, the chloride-secreting teleost opercular epithelium; ref. 13), one ion accounts for all of the I_{sc}. For cat trachea, the I_{sc} is a mixture of chloride secretion and sodium absorption (14). For large intestine, chloride secretion, sodium absorption, and potassium secretion or absorption can contribute to the I_{sc} (15).

Protocol 5. Determining unidirectional isotopic $^{36}Cl^-$ fluxes across an epithelium mounted in an Ussing chamber

1. Purchase isotope as $H^{36}Cl$.
2. Neutralize isotope with NaOH. Use data on isotope specification sheet to calculate amount of NaOH needed to neutralize.
3. Dilute in 1 or 2 ml of Ringer's solution.
4. Mount tissue in Ussing chamber and allow it to come to a steady state while short circuited. To calculate transepithelial electrical resistance (use procedure in Section 2.1.1), take electrical potential difference and I_{sc} measurements before the start of the experiment and at each sample time. Do not pulse the tissue to determine the resistance.
5. After removing an equivalent volume, add 25–50 μl of the isotope stock solution to achieve a final 5 μCi of radioactivity in the Ringer's bathing one side of the epithelium.
6. Determine c.p.m. in removed volume, using scintillation counter (= background).
7. Allow system to equilibrate for 1 h to ensure a linear flux of isotope during the cold side sampling.
8. After this initial hour, take 100 μl samples of Ringer's solution from the cold side every 30 min and 25 μl samples from the hot side every 60 min. Samples taken from the cold side should be replaced with equal volumes of non-isotopic Ringer's solution to keep the fluid volume constant on that side of the chamber. Take short-circuit current and potential difference measurements right before taking the sample.
9. Dissolve samples in a scintillant and count in a liquid scintillation counter.
10. Determine the average unidirectional fluxes from a minimum of four half-hour flux periods (see Section 7.2 below).

7.2 Sample calculation of isotopic flux from *Protocol 5* data

Data from the scintillation counter will include the background and isotope counts of the hot side and the cold side. The isotope counts must be corrected for background. Each data point can be converted from the scintillation data to an amount of $^{36}Cl^-$ in the chamber side at that sample time. There are two major complications in the final calculation of isotope flux. First, cold-side sampling removes counts which would have been counted in the following period had they not been removed. These counts must be added back. Second, since the goal is to know the amount of isotope that moves from the hot side to the cold side during a given time-period, it is necessary to subtract the number of counts that moved during the previous period.

As a specific example, consider conditions in which the cold-side sample size is 100 μl and the chamber has a volume of 2500 μl. At the end of the equilibration period 100 c.p.m. are detected (after background is subtracted; background will be subtracted for subsequent calculations in this example). Multiply this by 25 to obtain the counts in 2500 μl (i.e. 100 × 25 = 2500 c.p.m.). After 30 more min we detect 200 c.p.m. in the 100 μl sample. Multiply this by 25, and determine that there are 5000 c.p.m. in the chamber volume. However, 100 c.p.m. of counts were removed in our 100 μl sample following the last period. These must be *added* to the c.p.m., i.e. 5000 + 100 = 5100 c.p.m. Now, to calculate the actual number of c.p.m. moved exclusively in the previous 30 min, the counts moved in the previous period (2500 c.p.m.) must be subtracted. The result is 2600 (5100 − 2500 = 2600) c.p.m. In other words, 2600 c.p.m. were actually transported during this last 30-min period. After 30 more min we detect 300 c.p.m. Multiply this by 25 and obtain 7500 c.p.m., then *add* 200 c.p.m. removed with the previous 100 μl sample (7500 + 200 = 7700 c.p.m.), then *subtract* the counts moved in the previous period to derive the actual counts moved in this last period (7700 − 5100 = 2600 c.p.m.). This is the basic strategy to derive c.p.m. moved across the epithelium during the 30-min periods.

The average c.p.m. moved in the example above averaged 2600 c.p.m. (2600 + 2600/2 = 2600). A conversion factor (see Section 7.3) is used to convert c.p.m./30 min to $\mu Eq/cm^2/h$ of $^{36}Cl^-$ flux. The measured isotopic flux, calculated by multiplying the average c.p.m. moved (2600) by the conversion factor (0.003224), is 8.38 $\mu Eq/cm^2/h$. Multiply this number by 26.8 to convert to $\mu amp/cm^2$ (see Section 7.4). The result is 224.6 $\mu amp/cm^2$. This technique permits the determination of a unidirectional flux. The investigator would determine two unidirectional fluxes, one from the serosal side to the mucosal side (efflux), and a second from the mucosal side to the serosal side (influx). The net flux is the difference between the efflux and influx of the isotope. If the unidirectional fluxes are not statistically different, then that ion is passively distributed across the epithelium.

7.3 Calculation of the conversion factor

This calculated number converts c.p.m. detected on the cold side of the epithelium to $\mu Eq/cm^2/h$. To calculate this number, one multiplies the c.p.m. by 2 (to calculate a whole hour's worth of flux from two 30-min samples), divides by the surface area (in square centimetres), and divides finally by the specific activity (in c.p.m./μEq). The general equation is:

$$\frac{c.p.m. \times 2/\text{surface area}}{\text{specific activity}}.$$

An example is shown for an Ussing chamber with an aperture of 3 mm (0.07 cm² area), with sampling done every 30 min, and a specific activity (as calculated in Section 7.5) of 8861.7 c.p.m./μEq.

$$\text{Conversion factor} = \frac{c.p.m. \times 2/0.07 \text{ cm}^2}{8861.7 \text{ c.p.m.}/\mu Eq} = 0.003224$$

The units for this factor are $\mu Eq/c.p.m. \times cm^2 \times h$

7.4 Conversion of $\mu Eq/h/cm^2$ to $\mu amp/cm^2$

One ampere = 1 coulomb/second. Faraday's constant is 96 490 coulomb/mole of Equivalent (Eq). Therefore 1 coulomb = 10.363768 μEq. An example is shown using these numbers to convert $\mu amp/cm^2$ to $\mu Eq/cm^2/h$.

26.8 $\mu amp/cm^2$ = 26.8 × 10⁻⁶ coulomb/sec × 1/cm² × 10.363768 μEq/coulomb × 3600 sec/h = 1.0 $\mu Eq/cm^2/h$.

Therefore, data observed as $\mu Eq/cm^2/h$ can be multiplied by 26.8 to convert to $\mu amp/cm^2$.

7.5 Calculation of specific activity

The specific activity is the c.p.m./μEq of the isotope. An example is shown using results from a typical experiment.
  Average of two samples of diluted isotope:

63178 c.p.m./50 μl of samples (average of two samples)
−39 c.p.m. background counts (Ringer's solution without isotope)
63139 c.p.m./50 μl

or 1262780 c.p.m./ml.
Total Cl⁻ measured with a chloridometer:
142.5 mEq/litre or 142.5 μEq/ml.
Finally, 1262780/142.5 μEq/ml = 8861.7 c.p.m./μEq.

8. Electrophysiological studies of the paracellular pathway

As detailed above, the Ussing chamber, used under short-circuit current conditions, can reveal actively and passively transported species. In certain chloride-secreting opercular and jaw epithelium of teleost, the chloride ion is actively transported (13, 16, 17) and the sodium ion moves passively (5, 16, 17). It is generally assumed that the passive sodium movement in these and other chloride-secreting epithelia occurs through the paracellular pathway, i.e. the tight junction (18). The passive nature of this sodium movement can be studied with several additional electrophysiological techniques that are briefly described here. Additional information can be obtained from the cited references.

8.1 Isotope fluxes measured under open-circuit conditions

Isotope fluxes can also be determined in an Ussing chamber under open-circuit conditions. The flux data from these measurements is subjected to a flux ratio analysis, which predicts the ionic flux of an ion at a given trans-epithelial potential difference. Analysis of the sodium flux ratio in the opercular epithelium has been conducted by Degnan and Zadunaisky (19). There was no significant difference between the observed ratio of 1.14 and the predicted value of 0.94, indicating that sodium behaves passively across this epithelium. Likewise, in the isolated skin of *Gillichthys*, the sodium flux ratio under open-circuit conditions does not differ from that predicted for a passive, independently moving ion (20). The chloride ion is actively transported across the opercular epithelium (5, 13). As shown by Degnan and Zadunaisky (20), the mean predicted and observed chloride flux ratios were significantly different.

8.2 Isotope fluxes measured at pre-selected voltages

In a second approach, an Ussing chamber-mounted epithelium can be clamped to preselected voltages and ion fluxes measured. Again, measured flux rates are compared with predicted values. In the killifish opercular epithelium, measured and predicted flux rates for sodium did not differ when clamped to zero, or ± 25 mV (21). These data for a chloride secreting epithelium suggest that sodium moves through a single, rate limitng barrier, presumably the paracellular conductive pathway.

8.3 Application of the vibrating probe

The vibrating probe can be used to scan the surface of a short-circuited epithelium to detect the location of current flow (22). Foskett and Scheffey (23) and Foskett *et al.* (24) have applied this instrument to study ion pathways in the teleost chloride cell of tilapia opercular epithelium. The opercular epithelium is a heterogeneous epithelium with scattered chloride cells and thus is ideal for the application of the vibrating probe. Chloride cells proved to be the only significant and conductive elements in this epithelium (25). Virtually all chloride cells exhibited short-circuit currents, and their total currents accounted for all of the tissue short-circuit current. The most important finding with regard to junctional permeablity was the comparison of total tissue conductance with that of chloride cell permeability. Chloride cell (both transcellular and junctional pathways) accounted for all but 0.5 mS/cm^2 of the total tissue conductance, with leak pathways near the edge of the tissue accounting for the remainder. These vibrating probe findings are very significant since they help limit the possible sources of passive sodium movement to structures at the chloride cells, and not at other cells in this heterogeneous epithelium.

8.4 Flux measurements for non-ionic solutes

Under certain conditions it is useful to assess passive transepithelial permeability of non-transported hydrophilic solutes through the tight junction (26). These fluxes can be done under short-circuit conditions, open circuit conditions, or under special conditions; for example, when osmotic gradients are established across the epithelium. Techniques are similar to those described for the flux of an ion, except that, of course, there is no need to calculate the current carried by the permeable solute. One interesting application of this general is the measurement of mannitol flux across T_{84} monolayers during the transmigration of polymorphonuclear leukocytes. It was clearly shown that mannitol fluxes increased during this transmigration (27).

Acknowledgements

I would like to thank Mike Kennedy, Kyle Suggs, Andrew McGraw, David Nelson, and Jim Stidham for helpful comments and Marion Hinson for helping prepare the manuscript. This is publication number 103 of the Grice Biological Laboratory of the College of Charleston, Charleston, SC. Supported by NIH (GM24766, GM29099), NSF (DCB-8409165), Cystic Fibrosis Foundation, American Heart Association (South Carolina Affiliate) grants, and an Established Investigatorship from the American Heart Association.

References

1. Cereijido, M. (ed.) (1991). *The Tight Junction*. CRC Press, Boca Raton, Florida.
2. Karnaky, K. J., Jr. (1991). In *The Tight Junction*. (ed. M. Cereijido) p. 175. CRC Press, Boca Raton, Florida.
3. Ussing, H. H. and Zerahn, K. (1951). *Acta Physiol. Scand.*, **23**, 110.

4. Oakley, B. and Schafer, R. (1978). *Experimental Neurobiology: A Laboratory Manual.* University of Michigan Press, Ann Arbor.
5. Degnan, K. J., Karnaky, K. J., Jr., and Zadunaisky, J. A. (1977). *J. Physiol. (London),* **271,** 155.
6. Civan, M. M. and Garty, H. (1990). In *Methods in Enzymology,* Vol. 192 (ed. S. Fleischer and B. Fleischer), p. 683. Academic Press, London and New York.
7. Dawson, D. C. (1991). In *Handbook of Physiology. Section 6. The Gastrointestinal System* (ed. M. Field and R. F. Frizzell), p. 1. American Physiological Society, Bethesda, Maryland.
8. Kidder, G. W. III (1973). *J. Biol. Phys.,* **1,** 143.
9. Schultz, S. G. (1980). *Basic Principles of Membrane Transport.* Cambridge University Press, Cambridge.
10. Reuss, L. (1991). In *The Tight Junction* (ed. M. Cereijido) p. 49. CRC Press, Boca Raton, Florida.
11. Valentich, J. D. (1991). *J. Tiss. Cult. Meth.,* **13,** 149.
12. Dharmsathaphorn, K. and Madara, J. L. (1990). In *Methods in Enzymology,* Vol. 192 (ed. S. Fleischer and B. Fleischer), p. 354. Academic Press, London and New York.
13. Karnaky, K. J., Degnan, K. J., and Zadunaisky, J. A. (1977). *Science,* **195,** 203.
14. Corrales, R. J., Coleman, D. L., Jacoby, D. B., Leikauf, G. D., Hahn, H. L., Nadel, J. A., and Widdicombe, J. H. (1986). *J. Appl. Physiol.,* **61,** 1065.
15. Hegel, V. and Fromm, M. (1990). In *Methods in Enzymology,* Vol. 192 (ed. S. Fleischer and B. Fleischer), p. 459. Academic Press, London and New York.
16. Marshall, W. S. and Bern, H. A. (1980). *Epithelial Transport in Lower Vertebrates* (ed. B. Lahlou), p. 337. Cambridge University Press, Cambridge.
17. Foskett, J. K., Logsdon, C. D., Turner, T., Machen, T. E., and Bern, H. A. (1981). *J. Exp. Biol.,* **93,** 209.
18. Frizzell, R. A., Field, M., and Schultz, S. G. (1979). *Am. J. Physiol.,* **236,** F1.
19. Degnan, K. J. and Zadunaisky, J. A. (1979). *J. Physiol. (London),* **294,** 483.
20. Marshall, W. S. (1981). *J. Physiol. (London),* **319,** 165.
21. Degnan, K. J. and Zadunaisky, J. A. (1980). *J. Membr. Biol.,* **55,** 175.
22. Foskett, J. K. and Scheffey, C. (1989). In *Methods in Enzymology,* Vol. 171 (ed. S. Fleischer and B. Fleischer), p. 792. Academic Press, London and New York.
23. Foskett, J. K. and Scheffey, C. (1982). *Science,* **215,** 164.
24. Foskett, J. K., Machen, T. E., and Bern, H. A. (1982). *Am. J. Physiol.,* **242,** R380.
25. Foskett, J. K., Bern, H. A., Machen, T. E., and Conner, M. (1983). *J. Exp. Biol.,* **106,** 255.
26. Madara, J. L. and Hecht, G. (1989). *Functional Epithelial Cells in Culture* (ed. K. S. Matlin and J. D. Valentich), Vol. 8, p. 131. Alan R. Liss, New York.
27. Nash, S., Stafford, J., and Madara, J. L. (1987). *J. Clin. Invest.,* **80,** 1104.

12

Compartmented culture analysis of nerve growth

ROBERT B. CAMPENOT

1. Introduction

The compartmented culture was initially designed as a model system to study the formation and maintenance of neuronal projections. In compartmented cultures, axons (neurites) originating from neurones plated in a proximal compartment grow across a silicone grease barrier and enter into a separate fluid environment within a distal compartment, analogous to axons from neuronal cell bodies in one location in the nervous system projecting their axons through nerves or tracts to targets in another location (*Figure 1*). The first experimental use of the system was to show that axons of sympathetic neurones will enter into and grow within a distal compartment only if it is supplied with nerve growth factor (NGF) (1). Among the important principles established by this work is that NGF promotes growth of the neurite at its site of application to the neurite and not elsewhere, which is exactly what would be expected of a factor whose function is to direct nerve growth to or within a target location. Thus, NGF locally produced *in vivo* within the target tissue (and by Schwann cells and fibroblasts in the nerve after injury) would be expected to locally promote nerve growth (and regeneration). Work with compartmented cultures indicates that the immediate mechanisms by which NGF promotes growth are localized to the site of growth and do not involve the cell body or gene expression.

Although sympathetic neurones have been the cell type most commonly studied in compartmented cultures, any culturable neurone type that produces a neurite several millimetres long could, in principle, be employed in this system. Also, manipulations need not be limited to the chemical environment, and the compartmented culture model can be used to investigate the interactions of neurites with cultured target cells or Schwann cells (2).

Distinct from its use as a model for the formation and maintenance of neuronal projections, the compartmented culture offers a host of experimental advantages such as allowing convenient measurement of neurite elongation rate and separate collection and analyses of the cell components in different

Compartmented culture analysis of nerve growth

a. Compartmented culture

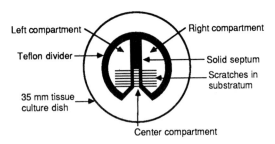

b. Enlargement of a single track with neurons plated in the center compartment

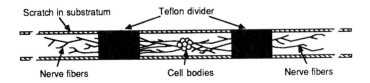

c. Enlargement of a single track with neurons plated in the left compartment

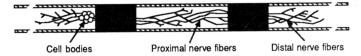

Figure 1. Schematic of a compartmented culture. (a) illustrates an entire culture, and (b) is an enlargement of a single track in a culture with neurones plated in the centre compartment. Neurites extend to the left and right under silicone grease barriers and into the separate fluid environments of left and right compartments. (c) is an enlargement of a single track with neurones plated in the left compartment. Neurites cross under a silicone grease barrier, span the centre compartment, cross under a second barrier and extend within the right compartment. The tracks are formed on the collagen-coated floor of a 35-mm plastic tissue-culture dish between a series of parallel scratches from which the dried collagen substratum has been scraped away. Each track is about 200 μm wide, the centre compartments are about 1 mm wide, and the barriers are about 0.5 mm wide. One culture can contain up to 20 tracks occupied by neurones.

compartments. A list of some of the experimental capabilities of the compartmented culture is given in the next section.

1.1 Experimental capabilities of compartmented cultures

Among the numerous experimental advantages of compartmented cultures are:

(a) The growth factors or other ingredients that are locally required for neurites to enter and grow within a distal compartment can be determined.

(b) Distal neurites can be maintained in a minimal medium (for example, without serum).

(c) The progress of neurites as they advance along the tracks can be measured quickly and non-invasively by means of a microscope digitizer that monitors stage position.

(d) Each culture contains about 16 usable tracks, each with its own group of neurones, so that it is easy to achieve an adequate sample of growth measurements.

(e) Distal neurites can be mechanically removed (neuritotomy) and reliably regenerate, allowing experiments to be initiated using cultures from maintained stocks rather than preparing a fresh plating of primary neurones each time.

(f) Cultures can be repeatedly neuritotomized, and the regeneration of neurites from the same neurones can be observed sequentially under different conditions.

(g) The effects of growth factors and pharmacological agents upon neurite growth can be investigated by applying the agents to the entire neurone, only to the cell bodies and proximal neurite segments, or only to distal neurites. This allows information to be obtained concerning the site of action of the factors and agents, as well as their functional specificity.

(h) Neurites in one distal compartment can be exposed to a treatment while neurites in the opposite distal compartment from the same population of neurones serve as controls.

(i) Metabolic labelling experiments can be performed in which either only cell bodies and proximal neurites or only distal neurites are exposed to a labelled precursor, and then the cell bodies and proximal neurites can be collected separately from the distal neurites for analysis. In this way it can be determined if products are made locally in distal neurites.

(j) Since the silicone grease penetrated by the neurites forms a restricted extracellular space, small electrical current pulses delivered between electrodes in adjacent compartments stimulate the crossing neurites to produce action potentials. Thus, the electrical activity of the neurones can be controlled while they grow and develop in culture, and whole populations of neurones can be electrically stimulated in physiological experiments.

2. Construction of compartmented cultures

The key to construction of compartmented cultures is the application of silicone grease to a Teflon divider and then seating the divider into a culture dish with a properly prepared substratum. To establish the goal at the outset, *Protocol 1* lays out this assembly procedure, and *Protocol 2* lays out the plating of neurones into compartmented cultures.

Before cultures can be constructed, a culture substratum must be suitably prepared (*Protocol 3*), requiring a pin rake which can be constructed in the lab (*Protocol 4*). Properly cleaned and sterilized (*Protocol 5*) Teflon dividers must be obtained (Section 2.5), as well as a syringe filled with sterile silicone grease (*Protocol 6*). The construction and uses of compartmented cultures has been previously described in briefer form (3).

2.1 Standard cultures

A standard compartmented culture is illustrated in *Figure 1a* and *b*. The first step in construction is the production of a suitable substratum which will orient the growth of neurites towards the distal compartments. The substratum most commonly used is collagen; the dish floors are first coated with a dried collagen film (*Protocol 3*), and tracks are formed by making a series of parallel scratches, in the dish floor (*Figure 2a*) with a single stroke of a specially constructed pin rake (*Protocol 4, Figures 5* and *6*). Bare plastic is exposed in the scratches and subsequent neurite growth will be confined to the collagen-coated plastic between them. Before seating the divider, a drop of methylcellulose-thickened culture medium is placed on the scratched substratum to wet this region (*Figure 2b*).

A sterile divider (*Protocol 5*) is then picked up by its solid septum with a 90 degree haemostat, oriented such that the divider is horizontal when the haemostat is placed upside down on the work surface under a stereomicroscope. Under the microscope, sterile silicone grease is applied to the seating face of the divider with a syringe with an 18 gauge needle squared off at the tip (*Protocol 6, Figures 4* and *7*) in a manner analogous to applying decorative cake icing. Then a culture dish is picked up with fingers, its lid is removed, and it is inverted, quickly so the drop of culture medium does not move or spread. The dish is then oriented over the divider and dropped into place (*Figures 2c* and *3*). Using fine forceps or other pointed object, the dish floor is gently pressed in the region of the perimeter of the divider and solid septum to seal the grease. The dish is not pressed in the scratched region, and because this region had been previously wetted, the silicone grease is prevented from directly adhering to the collagen substratum. A gasket-like seal results which is sufficiently tight to prevent bulk flow of medium or significant diffusion of medium components between compartments (see Section 3.3), but is penetrable by growing neurites. After seating the divider, the divider/dish assembly is inverted and released from the haemostat so that it drops gently to the work surface. A small dab of silicone grease is placed on the substratum at the mouth of the centre compartment (*Figure 2c*) to prevent the cell suspension that will be injected there from flowing out, and the lid is replaced on the dish. Once assembled, the dishes are placed in the incubator for one or more hours to allow the grease to flow slightly and the divider to settle. Then the side compartments are filled with culture medium. If the side

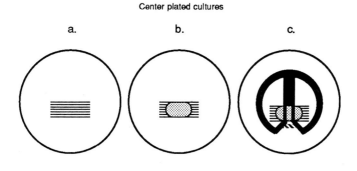

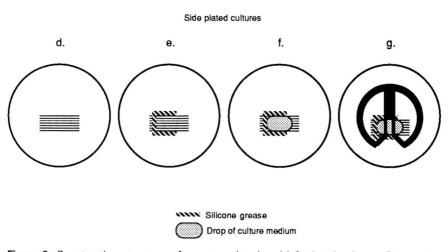

~~~ Silicone grease
◯ Drop of culture medium

Figure 2. Construction sequences for centre-plated and left-plated cultures. For centre-plated cultures a 35-mm tissue culture dish is coated with collagen and scratched to form tracks, illustrated in (**a**). A drop of culture medium has been placed on the scratched substratum in (**b**). In (**c**) the Teflon divider has been seated and a dab of silicone grease has been placed at the mouth of the centre slot. For left-plated cultures a dish is similarly prepared (**d**), but a C-shaped 'rope' of silicone grease is applied to the substratum as shown in (**e**). Then a drop of culture medium is placed on the scratched substratum (**f**) and the Teflon divider is seated in place (**g**).

compartments are filled with culture medium immediately after dish assembly without the settling period, leaks are frequent.

2.2 Side-plated cultures

For certain types of experiments it is advantageous to plate the neurones in a side compartment and have their neurites grow across the centre compartment and into the opposite side compartment. By convention, we plate

Compartmented culture analysis of nerve growth

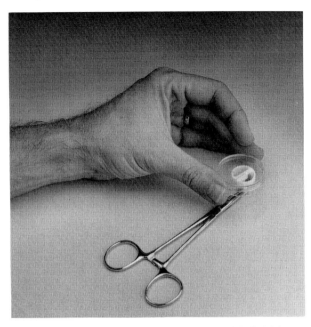

Figure 3. Photograph showing a 35-mm tissue culture dish held inverted above the greased divider just prior to seating it in place.

neurones in the left compartment, and the procedure for constructing left-plated dishes is similar to centre-plated dishes, except that a well must be formed in the left compartment to contain the suspension of neurones in the scratched region. This is accomplished by laying down a C-shaped perimeter of silicone grease in the scratched region of the dish floor as indicated in *Figure 2e* before the region is wetted with culture medium. The grease is allowed to settle for about 1 h in the incubator. A drop of methylcellulose-thickened culture medium is then placed within the C (*Figure 2f*), silicone grease is applied to the divider exactly as with centre-plated cultures, and the dish is seated on the divider as indicated in *Figure 2g*. The divider is allowed to settle in the incubator for one or more hours, and then the centre and right compartments are filled with culture medium. Ordinarily, the medium pipetted into the outer perimeter of the dish will not flow into the centre slot because of the hydrophobicity of the Teflon and silicone grease. It is usually necessary to inject medium into the centre slot with a syringe and needle in order to fill the slot and to establish confluence between the medium in the slot and the medium in the outer perimeter.

Since the walls of the well which will contain the neurones are low on three sides, consisting only of the silicone grease, the well can hold only a small volume of culture medium. Therefore, before plating the neurones, any residual medium remaining within the silicone grease perimeter in the left compartment (*Figure 2g*) is removed with a syringe and needle. A suspension of cells in methylcellulose-thickened medium is then injected into the well, taking care to ensure that the suspension adheres as well as possible to the Teflon divider and the meniscus stands as high as possible without spilling out of the well. This maximizes the volume of suspension contained within the well. The next day after the cells have settled on the substratum, the left compartments are topped up with culture medium. As with centre-plated cultures, the neurones are allowed to settle overnight, and then the left compartments are filled with culture medium. After about 1 week, neurites have crossed both barriers and are well-established in the right compartments.

Protocol 1. Assembling the dish and divider

Materials
- collagen coated and scratched tissue culture dishes (*Protocol 3*)
- culture medium
- tray for 12 cultures made from black Plexiglass (optional)
- Petri dish containing sterile Teflon dividers (*Protocol 5*)
- haemostat with 90-degree angled jaws
- stereo-microscope set up in a sterile hood
- two pairs of fine forceps (not dissection quality)
- two sterile silicone grease syringes (*Protocol 6*)
- sterile, disposable pipettes
- pipette bulbs
- burner

Procedure
1. For left-plated cultures skip this step and proceed to step 2. For centre-plated cultures arrange 12 culture dishes on the black Plexiglass sheet in a sterile hood equipped with a stereo-microscope. Under the stereo-microscope place a drop of methylcellulose-containing culture medium on the scratched region of the floor of each dish (*Figure 2b*). Take care not to touch the pipette to the substratum, and keep the drop confined within the width of the scratched area. Set the dishes aside. Proceed to step 3.
2. For left-plated cultures arrange 12 culture dishes on the black Plexiglass sheet in a sterile hood equipped with a stereo-microscope. Under the stereo-microscope lay down a C-shaped perimeter of silicone grease in each dish as shown in *Figure 2e*. Set the dishes in the incubator to allow the grease to settle for about 1 h. After the grease perimeter has settled, place a drop of methylcellulose-containing culture medium on the scratched

Protocol 1. *Continued*

region of the floor of each dish within the C (*Figure 2f*). Take care not to touch the pipette to the substratum. Set the dishes aside.

3. Flame-sterilize the haemostat. Remove a Teflon divider from the Petri dish by gripping the rim of the divider with sterile forceps. Transfer the divider to the haemostat, clamping the divider by the solid septum. Lie the haemostat on its back on the work surface such that the bottom of the divider is visualized under the stereo-microscope. Apply silicone grease to the divider, taking care that the regions under which the neurites will cross receive a neat 'rope' of grease (*Figure 4*).

4. Pick up a culture dish, set aside its lid, and invert the dish quickly so as not to disturb the drop of medium. Bring the dish into position over the divider and drop it in place (*Figures 2c* and *3* for centre-plated cultures, *Figures 2g* and *3* for left-plated cultures). With fine forceps press down on the dish in the area of the solid septum and around the rim to seat the grease. Make a mental note of the position of any gaps in the grease around the perimeter. Do not press in the region where the neurites will cross. When the dish is properly seated (a judgement developed with experience), pick up the haemostat, turn it over, and release the dish so that it falls gently on the work surface (placing it on the surface can disrupt the seating of the divider). Calk any regions around the perimeter of the divider where gaps in the grease were visible. Only for centre-plated cultures, place a dab of grease at the entry of the centre slot (*Figure 2c*). Replace the cover of the dish.

5. Repeat steps 3 and 4 for each dish, and then store the tray in the incubator. After 1 or more h in the incubator fill the compartments which will not be receiving neurones with culture medium; i.e. left and right compartments in centre-plated cultures, centre and right compartments in left-plated cultures.

2.3 Plating the neurones

Neurones are usually plated the day after the culture dishes were constructed. Dissociated neurones, typically sympathetic neurones from newborn rats, are suspended in methylcellulose-thickened medium and loaded into a disposable sterile syringe fitted with a 1.5 inch, 22-gauge needle (*Protocol 2*). The high viscosity of the medium prevents the cells from settling in the syringe during the procedure. Cell suspension is injected into the centre slots of each compartmented dish under the dissecting microscope. The slot is carefully filled to the top, and the hydrophobicity of the Teflon and the dab of silicone grease on the substratum at the mouth of the slot are sufficient to prevent the suspension from leaking out into the bulk of the dish. The neurones are allowed to settle on to the substratum overnight, and by morning many of the

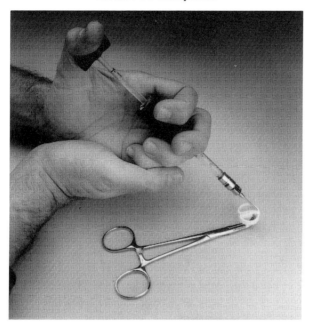

Figure 4. Photograph showing the application of silicone grease to a Teflon divider. The divider is held with a 90-degree haemostat resting on the work surface. This operation is performed under a stereo-microscope.

neurones will have sprouted neurites. At this time the outer perimeter of the dish is filled with culture medium (1–1.5 ml), making sure that it establishes confluence with the centre slot medium. By the next day neurites begin to emerge into left and right compartments.

Protocol 2. Plating the neurones

Materials

- cell suspension
- 1 ml sterile disposable syringe with 22-gauge, 1.5-inch needle (two syringes needed for left-plated cultures)
- sterile, disposable pipettes
- pipette bulbs

Procedure

1. Obtain a cell suspension of dissociated sympathetic neurones as previously described (4), modified from Hawrot and Patterson (5). The neurones from 20 ganglia are suspended in about 1.3 ml of L15 CO_2 medium containing methylcellulose, nerve growth factor, rat serum, cytosine arab-

Protocol 2. Continued

inoside, and vitamin C. For centre-plated cultures use 0.8 ml cell suspension/12 dishes, and for left-plated cultures, use 0.5 ml/12 dishes. Allow 0.5 ml waste.

2. Transfer the cell suspension to a 60-mm plastic Petri dish, propped in a tilted orientation such that the fluid remains on one side, and load the cell suspension into a 1-ml syringe fitted with a 22-gauge, 1.5-inch needle.
3. For centre-plated cultures, inject cell suspension into the centre slot of each dish, filling the slot.
4. For left-plated cultures, remove residual medium from within the silicone grease perimeter in the left compartment which will contain the neurones (*Figure 2g*) with an empty 1-ml syringe and needle, and then inject the cell suspension with care so that it adheres to the Teflon divider.
5. The next day, after the cells have settled on the substratum, top up the cell-containing compartments with culture medium.

Protocol 3. Preparation of collagen substratum

Materials
- sterile plastic tissue culture dishes (35 mm diameter)
- fresh, sterile rat tail collagen solution
- sterile, tissue culture-quality distilled water
- pin rake (*Protocol 4*)
- tray for 12 cultures made from black Plexiglass (optional)
- sterile, disposable pipettes
- pipette bulbs

Procedure

1. Prepare a solution of rat tail collagen as described by Hawrot and Patterson (4) using sterile technique (see also Chapter 9). The shelf-life of collagen solution used for this purpose is 2 weeks in the refrigerator. Collagen-coated culture dishes may be stored frozen indefinitely. 'Bad' collagen will peal from tracks during scratching, producing an excessive amount of white powder visible to the naked eye.
2. To coat culture dishes, dilute the collagen solution (4 parts sterile, culture-quality distilled water to 1 part collagen solution for sympathetic neurone cultures; 1 part water to 1 part collagen solution for dorsal root ganglion neurone cultures). Wet the floors of 35-mm tissue-culture dishes by pouring enough solution into the first dish to fill it about half and then pouring the solution from the first dish into the second dish and so forth. Replenish

the solution as needed. This will leave behind a film sufficient to coat the floors of the dishes. Air-dry the dishes with lids on overnight in a laminar flow hood.

3. Form the collagen tracks by scoring the collagen-coated floors of the tissue culture dishes with a pin rake (*Figures 5* and *6*). Use sterile technique in a laminar flow hood. It is helpful to place the dish on a black Plexiglass background to permit visualization of the scratches as they are made. The tracks should be about 20 mm long, with the scratches positioned as indicated in *Figure 2a*. Store the dishes at $-20\,°C$ if storage will be for more than a day or two.

2.4 Pin rake

In order to construct compartmented cultures, a pin rake (*Figures 5* and *6*) is needed to make a series of parallel tracks in the substratum. Constructing one is a fussy project, but one can be successfully constructed in the laboratory by carefully following *Protocol 4*. It should be noted that, while a particular project may not require that the cultured neurones be organized on a substratum of parallel tracks, if tracks are omitted the neurites cross under the silicone grease relatively poorly. Neurites on a track commonly split into two fascicles which run under the silicone grease, hugging the scratches that

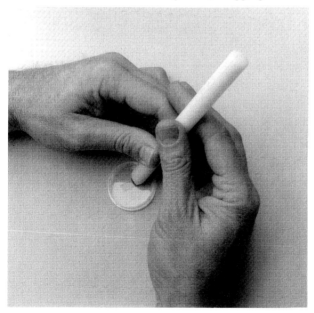

Figure 5. Photograph showing a 35-mm tissue culture dish being scratched with the pin rake. The rake is held as shown and the scratches made with a single stroke away from the body.

Figure 6. Pin rake. 21 insect pins are cemented together with 5-min. epoxy and mounted in a handle made from a length of rod or tubing. Epoxy is built up to form a cone from the handle nearly to the tips of the pins as shown to provide support, thereby preventing the pins from bending during the scratching procedure.

border the track. Thus, it seems likely that there is resistance to crossing under the grease that causes the pioneer growth cones to turn away until they are trapped in the corner formed by the grease and the scratch. Apparently this forces them to penetrate the grease. Once pioneers have crossed it is presumably easier for subsequent growth cones to follow the same path, producing the observed fascicles.

Protocol 4. Construction of the pin rake

Materials
- 21 insect pins (size 00)
- dissecting microscope
- hot plate
- aluminium sheet (about 3 inch × 3 inch × ½ inch)
- Parafilm
- fine forceps
- quick-setting (5-min) epoxy cement
- phenolic or Plexiglass rod $^{7}/_{16}$-inch diameter × 6 inch
- small handsaw

Procedure

1. A pin rake (*Figure 6*) can, in principle, be constructed from any type of pins. We typically use size 00 insect pins, which are about 200 μm in diameter and produce collagen tracks about 200 μm wide.

2. Cut the heads off of 21 of the size 00 insect pins.

3. Arrange a hot plate so that its surface may be viewed with a dissecting microscope. Place a piece of thick aluminium sheet (about 3 × 3 × ½ inch) on the hot plate, and melt a piece of Parafilm (about 2 inch by 2 inch) on to the surface of the aluminium sheet to produce a sticky (not liquid) surface. Position the Parafilm such that it comes up to one of the edges of the aluminium sheet.

4. Holding the back end of one of the pins with fine forceps, embed a pin in the Parafilm such that the point is visible under the dissecting microscope and the back end extends over the edge of the aluminium plate. The pin may be pressed into the Parafilm with the side of the forceps. Place the next pin similarly, locating it alongside first pin with the points as even as possible. Continue placing pins until there are 21 pins, side by side, with their points as even as possible. Then, using the back end of the forceps as a straight edge, push against the points of the pins to move them slightly backward to even out the line of the points. In this way the points can be made to line up almost perfectly. Turn off the hot plate.

5. Mix some 5-min epoxy cement and put several drops on the pins at about their midsection to cement them together. After the epoxy has set, lift the pin rake from the Parafilm.

6. A handle for the pin rake can be made from Plexiglass or phenolic tubing or rod of about $7/_{16}$-inch diameter. A slot to accommodate the pins must be sawn in the end of the rod/tube across the diameter and about 1.25-inch along the length. The back end of the pins are then placed in the slot such that the points extend about ½ inch beyond the handle. Once positioned, the pins are glued in place with epoxy and, in a series of applications of epoxy, a cone is built up around the pins such that only about ⅛ inch of their ends extend beyond the epoxy.

7. Clean and sterilize the pin rake by wiping with a Kimwipe soaked in 70% ethanol and then allowing it to air-dry in the hood.

2.5 Teflon dividers

The most essential and difficult item to obtain is the Teflon divider. These are now commercially available in virtually any custom design (see below). While they are not inexpensive, they are indefinitely reusable, so divider availability is not a serious barrier to establishing the compartmented culture technique in the laboratory.

A variety of compartment configurations have been used, but the most common type of divider is illustrated in *Figure 1*. This style of divider can be machined from ¾-inch Teflon rod stock and divides a 35-mm tissue culture dish into left and right compartments separated by a septum into which the centre compartment is machined. The centre compartment consists of a slot extending from the mid-point of the septum to the perimeter of the divider where it communicates with the bulk of the dish. The only critical dimensions are the slot width which is 1 mm, and the walls separating the centre slot from the side compartments which are 0.8 mm thick. Chronic leaking is a problem if the walls are much thinner. The walls around the perimeter are thicker, about 1.5 mm. The divider is 6 mm high, which is high enough to hold 0.5 ml of medium in each side compartment, but low enough so that the medium in a full side compartment will not contact the dish lid. Dividers of this style have been made by a machine shop using end mills in a computer-controlled milling machine. Teflon dividers are reusable and can be autoclaved, but the first autoclaving often causes distortion of dividers machined from rod stock. After the first autoclaving they must be sanded flat by rubbing them on fine sandpaper (240 grit works well) on a flat bench top, and then the ragged edges must be carefully trimmed with a No. 11 scalpel. When sanding, a convenient way to hold the dividers is with a 90-degree haemostat clamped on the solid septum.

Recently dividers have been made by Tyler Research Instruments (Edmonton, AB, Canada) using a computer-controlled Paser milling machine which cuts with a high-velocity water jet. Dividers of any shape can be made from Teflon sheet without the constraints inherent in the circular cutting motion of conventional milling machines. Another advantage is that dividers machined from 3-mm or thicker sheet do not distort when autoclaved, so the sanding and trimming steps are unnecessary.

It should be pointed out that when the compartmented culture system was first developed, Teflon rings sliced from tubing were used to divide culture dishes into two compartments, and neurones were either plated inside the ring and extended their neurites to the outside, or were plated outside and extended their neurites inside. The neurones were unhealthy in appearance and results were poor using rings, which was first attributed to the possibility of toxic substances in the silicone grease, but in the end the problem was attributed to the hydrodynamic effect of the ring. Medium tended to swirl around the ring whenever the dishes were handled, which apparently caused physical stress on the cell bodies and/or neurites on the outside of the ring. Dividers which protect the neurones within a centre slot solved the problem. Thus, the standard divider design originated as a way of protecting the neurones from swirling medium. Potential users should be cautioned against the use of rings, and should take hydrodynamics into consideration when varying the divider design.

Protocol 5. Cleaning and sterilization of Teflon dividers

Materials
- Teflon dividers (Tyler Research Instruments)
- 70% ethanol

- two beakers of suitable size to hold the dividers
- 4-litre jug of sulfuric acid with Nochromix (Godax Laboratories, Inc., New York, NY)
- glass funnel to fit the sulfuric acid jug
- glass Petri dishes (10 cm)
- forceps (not dissection quatity)
- Kimwipes

Method

1. Remove the dividers from the culture dishes with forceps and wipe off the silicone grease with a Kimwipe. Place the dividers in a beaker containing 70% ethanol to await acid cleaning.
2. Pour off the ethanol, place the baker in a fume hood and add clean sulfuric acid with Nochromix. Leave until dividers appear pure white. (They can be left indefinitely in the acid without harm.)
3. Pour the acid back into its jug, catching the dividers in the funnel. Transfer them to a clean beaker. Rinse the dividers five times with culture-quality distilled water to remove the acid, and then boil them in the water for 1 h. Pour off the hot water and rinse again five times.
4. Use clean forceps to move the dividers to clean Petri dishes with as little water as possible. Autoclave them, and dry them in an oven if necessary.

2.6 Silicone grease syringe

Sterile silicone grease is applied through a syringe and needle (*Figures 4* and *7*), and while this would seem straightforward, there are some aspects worth describing. A 1-ml glass, luer-lock syringe should be used, and it should be padded to protect fingers from the high force required to squeeze the silicone grease through an 18-gauge needle (*Protocol 6*). Loading the syringe and sterilization are described in the protocol, but a very important point should

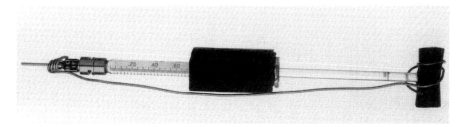

Figure 7. Silicone grease syringe. A silicone grease applicator is made from a 1 ml glass luer-lock syringe fitted with a 16-gauge needle with a squared-off tip. Rubber tubing is used to pad the plunger and form a grip on the barrel to aid in squeezing out the grease.

Compartmented culture analysis of nerve growth

be emphasized about withdrawing the plunger to reload the grease. The plunger should be withdrawn extremely slowly since the film of grease between the plunger and barrel can create enough resistance so that too much force on the plunger will shatter barrel, possibly causing injury.

Protocol 6. Preparation of silicone grease syringe

Materials
- Dow Corning high-vacuum grease in 150-g tube
- 1-ml glass luer-lock syringe
- 18-gauge hypodermic needle with a squared-off tip
- 2-inch length of vacuum tubing
- ¾ inch of ordinary rubber tubing
- stiff, uninsulated wire
- 10-ml disposable syringe
- pliers
- Kimwipes

Procedure

1. Silicone-grease is applied to the Teflon divider through a 1-ml glass, luer-lock syringe fitted with an 18-gauge needle with a squared off tip (*Figures 4 and 7*). A needle about 2 cm long works best. Because some force is required to squeeze the grease through the needle, it is beneficial to slip a 2-inch length of vacuum tubing over the barrel and to pad the plunger by taking a ¾-inch length of rubber tubing of suitable diameter, cutting a hole in the side of the tubing midway between the ends, and slipping the head of the plunger through the hole. In this way pressure can be applied with the first and second fingers crooked over the end of the vacuum tubing encasing the barrel and the thumb pressing against the padded plunger.

2. To fill the applicator syringe with silicone grease, first squeeze the grease from its tube into a 10-ml disposable syringe with no needle.

3. Remove the needle from the applicator syringe with pliers, and if there is grease remaining from previous applications, empty it by fully depressing the plunger. Then slowly (so as not to shatter the syringe barrel!) withdraw the plunger completely and set it aside.

4. Hold the 10-ml syringe by the body in one hand, nipple up with its plunger resting on the laboratory bench. Hold the body of the applicator syringe in the other hand and place it over the nipple of the 10-ml syringe, holding it to make a tight seal. Press down with both hands so as to depress the plunger of the 10-ml syringe and fill the 1-ml syringe about three-quarters

full with silicone grease. Then insert the plunger into the 1-ml syringe and depress it until grease comes out of the nipple. Wipe away grease with a Kimwipe. Install the needle and depress the plunger until grease comes out of the needle. Wipe away grease with a Kimwipe.

5. Autoclave the filled applicator syringe with attention to the following details: Any air bubbles trapped in the applicator syringe can expand during autoclaving and blow out the plunger. To prevent this, attach the plunger securely to the syringe by looping some ordinary stiff uninsulated wire over the end of the plunger and around the barrel of the syringe below the vacuum tubing. Bubbles will then harmlessly force grease out of the needle. Applicator syringes need not be wrapped for autoclaving but should be placed in a pan or on aluminium foil to catch the exuded grease. Store the filled applicator syringe until needed.

6. Before using the applicator syringe, squeeze out some grease, wipe it away with a Kimwipe, and then sterilize the exterior by dipping into 70% ethanol and allowing it to air-dry.

3. Culture medium

One of the principle uses of compartmented cultures is exposure of distal neurites to culture medium that is different from the medium supplied to cell bodies and proximal neurites. This is done not only as an experimental variable, but also as standard procedure to establish cultures with certain desirable characteristics. Several issues which have arisen concerning the composition of medium are discussed below.

3.1 Methylcellulose

When the compartmented culture was first developed, the culture medium used was a modified L15 medium, called L15 CO_2, which is described by Hawrot and Patterson (5). This medium contained methylcellulose, a thickener, to make it favourable for culturing pure cultures of sympathetic neurones which are not as firmly anchored to the collagen substratum as when they are cultured on a layer of non-neuronal cells. With methylcellulose the hydrodynamic stresses associated with handling the dishes and changing the medium are reduced, allowing the neurones to remain attached.

3.2 Rat serum and ascorbic acid

The medium also contained 2.5–5% rat serum, which was needed for neuronal survival, and 1 mg/ml ascorbic acid, a cofactor in noradrenaline synthesis which is included at such a high concentration because of its rapid breakdown. Experiments with compartmented cultures showed that neurites survive and grow in distal compartments containing medium without rat serum, the neurites spread out over and appeared to adhere well to the collagen substratum, and they reliably extended along the tracks at a rate of about 1 mm/day. In contrast, with rat serum the neurites were highly cabled, appearing to prefer to adhere to one another rather than to the substratum, the cables frequently detached from the substratum during medium changes, and the rate of neurite extension was low and variable (6). Because of the favourable growth characteristics, rat serum would have been thereafter omitted from the distal compartments had it not been for the observation that without serum the ends of the neurites degenerated when the medium was changed. In response to fresh, serum-free medium the distal 0.5 mm of the neurites degenerated. It is not known whether similar degeneration occurred in response to fresh rat serum-containing medium, since it may have escaped notice in the highly cabled, less regularly extending neurites. Experiments showed that the degeneration was not the result of mechanical damage, and either ageing the medium in the incubator for several days before giving it to the cultures or omitting the ascorbic acid abolished the damaging effect. Thus, it appeared that the high concentration of ascorbic acid present in fresh medium was damaging to the distal neurites (6). It has since become standard procedure to only supply rat serum or ascorbic acid to the centre compartments containing the cell bodies and proximal neurites, while distal neurites are given medium without these ingredients. Fresh medium containing rat serum and ascorbic acid given to cell bodies and proximal neurites produces no observed ill effects, and distal neurites extend after medium changes without degeneration or interruption.

NGF is routinely supplied in culture medium at 200 ng 2.5 S NGF/ml, and is required for the survival and growth of sympathetic neurones from superior cervical ganglia of newborn rats. However, once neurites have crossed from the centre compartment into the distal compartments, it is sufficient for neuronal survival to supply NGF only to distal neurites, and it is usually experimentally advantageous to discontinue supplying NGF in centre compartments containing the cell bodies and proximal neurites after about 6 days in culture. Discontinuation of NGF in centre compartments causes neurite growth there to virtually cease and subsequent neurite growth is confined to the distal compartments where it can be observed, analysed, and removed by neuritotomy (7). Cultures maintained with NGF supplied in the centre compartments typically display far less neurite growth in the distal compartments than cultures which have had NGF discontinued in the centre compartments. Also, if NGF is present in the centre compartment, a considerable proportion of the regeneration after neuritotomy will occur in the centre compartment where it cannot be easily observed or removed by subsequent neuritotomy. For similar reasons it is common practice to discontinue NGF in the left and centre compartments of left-plated cultures at about 1 week in culture after neurites are well-established in right compartments.

3.3 Effectiveness of the seal between compartments

A principal experimental use of compartmented cultures is exposure of distal neurites to different medium than supplied to cell bodies and proximal neurites. The barrier between compartments is not absolute, since a film of wetted collagen extends beneath the silicone grease, but the barrier effectively prevents the bulk flow of culture medium between compartments. Ordinarily, the side compartments are filled to the top and a relatively low level of medium is maintained in the centre slot and dish perimeter, and this condition persists over 6 days between medium changes indicating no gross leakage of medium. Leaks occasionally occur, and when cultures do leak, the medium in the side compartment and centre compartment equalizes over a relatively short time. When this happens cultures are eliminated from the experiment.

The silicone grease seal also prevents the inter-compartment diffusion of growth factors and small molecules such as ions which has been demonstrated in two ways. First, when NGF has been withdrawn from centre compartments, NGF in the distal compartments supports the survival of neurones on tracks that have neurites crossing into the distal compartments (1). But in many cultures there are one or two tracks at each edge of the scratched region which were not contacted by the drop of medium used to wet the collagen during dish assembly. The silicone grease is sealed to dry collagen on these tracks and neurites of neurones on these tracks cannot cross into side compartments. Neurones on sealed tracks die after NGF is withdrawn from the centre compartment, while neurones on wetted tracks adjacent to them survive by virtue their neurites crossing into NGF-supplied side compartments (unpublished results). The neurones on sealed tracks then serve as a bioassay indicating that NGF is not leaking from side compartments in amounts sufficient to support neuronal survival. In this experiment the fluid levels in the NGF-containing side compartments had been kept lower than in the centre compartment so that any imperceptible bulk flow of medium would serve to slightly dilute the NGF in the side compartments rather than carry NGF into the centre compartment.

Second, measurements of Ca^{2+} in medium showed that very little Ca^{2+} moved from compartments supplied with ordinary medium (1 mM Ca^{2+}) into compartments supplied with Ca^{2+}-free medium (4). Atomic emission measurements of total Ca^{2+} showed that nominally Ca^{2+}-free medium actually contained 10 µM Ca^{2+} (which was brought down to less than 0.5 nM by 5 mM EGTA). When this medium was supplied in side compartments of cultures from the time of plating, neurites entered and grew within the side compartments. After 6 days in culture the side compartment medium was found to contain about 20 µM Ca^{2+} (an additional 10 µM). Thus, a 100-fold Ca^{2+} concentration difference was maintained, and the seal was probably much more effective than that since much of the additional Ca^{2+} was likely to have been transported across and released by the neurites.

4. Applications of compartmented cultures
4.1 Neuritotomy and collection of neurites

One of the major advantages of compartmented cultures for studies of nerve growth arises because distal neurites reliably regenerate after they have been mechanically removed from distal compartments with a jet of fluid delivered with a syringe (*Protocol 7*). This removal, termed neuritotomy, is quick, simple, reliable, and it allows stocks of cultures of primary neurones to be maintained in the laboratory and used as needed without necessitating a fresh dissection of primary neurones for each experiment. Neuritotomy also allows the use of favourable experimental designs. For example, the regeneration of distal neurites from the same population of neurones can be observed sequentially under different experimental conditions (for example, growth factor concentration or presence or absence of a drug), so that the neurones serve as their own controls, and reversibility is easily determined. Another advantage is that neuritotomy establishes identical starting times for all the tracks of a culture, resulting in less track-to-track variability as the neurites grow along towards the ends of the tracks.

Neuritotomy is performed either for the purpose of observing subsequent regeneration, or for collecting the neurites for biochemical analyses, or both. When the purpose is only to observe subsequent regeneration, it is best to use distilled water as the wash fluid. This apparently combines osmotic with mechanical removal of distal neurites, since the use of isotonic saline is somewhat less effective, requiring a more vigorous jet to completely remove the distal neurites. However, isotonic saline works well enough and can be used when needed for collecting neurites for certain analyses. With these physical removal techniques, three successive washes are usually sufficient to remove all visible trace of neurites. If neurites are to be analysed, each wash is collected with a Pasteur pipette. If the cell bodies and proximal neurites in the centre slot are also to be collected for analysis, they can be similarly washed while the dish is tilted so that the washed-away cells are confined to dish perimeter near the centre slot where they can be easily collected. If side and centre compartments are washed with the divider in place, neurites under the silicone grease will be left behind. Alternatively the divider can be removed after the neurites in the left and right compartments have been collected, and then the entire dish can be washed to collect all the remaining cellular material.

Protocol 7. Neuritotomy and collection of cellular material for analysis

Materials
- sterile 3-ml disposable syringe with a 22-gauge 1.5-inch needle
- serum bottle containing sterile, culture quality distilled water (or container with an appropriate buffer for collecting neurites for analysis)

- culture medium of desired composition
- dissecting microscope set up in a sterile hood
- sterile pipettes for dispensing medium
- pipette bulbs
- sterile Pasteur pipette set up as an aspirator

Procedure

1. Load a sterile 3-ml syringe with needle with sterile water (or appropriate buffer).
2. Orient a compartmented culture under the dissecting microscope such that it is rotated 180 degrees from the ordinary orientation. Remove the lid and aspirate the medium from the left and right compartments.
3. Using the 3-ml syringe, direct a fairly vigorous jet of water (or buffer) at the substratum bearing the neurites until the compartments fill with fluid. Do not aim the jet directly at the silicone grease barrier. Aspirate (or collect) the water (buffer) and repeat two more times.
4. If the culture is to be maintained, supply the appropriate medium to the compartments, cover the dish, and return it to the incubator.
5. If the cell bodies and neurites in the centre compartment are to be harvested, remove and collect them by a similar washing procedure. Note that if it is desired to collect all of the cellular material, the divider should be removed before washing the cell bodies and proximal neurites so as to also collect the neurites that resided under the barrier.

4.2 Measurement of neurite extension

One of the major uses of compartmented cultures is the measurement of neurite growth. What is actually measured is the distance that neurites extend along each track starting from the edge of the silicone grease at the base of each track. Since the distances measured are greater than a microscopic field, measurements are made with an instrument attached to the microscope stage that monitors stage position. This can be accomplished simply and cheaply by mounting a digital machinist's dial gauge such that the plunger follows left-right (X-axis) stage movements, and then recording the measurements manually, but much faster measurement and analysis can be accomplished using an MD2 microscope digitizer (Minnesota Datametrics, St Paul, Minnesota). Optical readout devices monitor X and Y stage position, and positions are recorded by pressing a foot pedal. The data are fed into an IBM (or compatible) personal computer and processed with custom software designed for analysing compartmented culture data (by Minnesota Datametrics).

There is a special problem in viewing and accurately measuring neurites in compartmented cultures with phase-contrast microscopy because menisci

Compartmented culture analysis of nerve growth

associated with the divider distort the phase ring and destroy the contrast. This problem can be corrected by means of a modification of the phase condenser. For information contact the author.

The first step in measuring neurite extension in a culture is to align the X-axis of the culture with the X-axis of the microscope stage. This is accomplished by positioning the culture on the stage such that the left and right ends of the top scratch in the culture substratum can both be brought under the eyepiece crosshairs by moving the X-axis of the stage left and right without moving the Y-axis. With practice this can be quickly and easily accomplished.

Protocol 8. Alignment of microscope stage

1. Place the culture on a rotatable platform on the stage (the inserts that are standard on the stage of the Nikon Diaphot fit this purpose nicely) and move it into a position where the left end of the top scratch in the culture substratum is under the eyepiece cross-hairs.

2. Then move the stage by only moving along its X-axis all the way to the right end of the scratch. This will (unless one is extremely lucky) be impossible since the right end of the scratch will lie above or below the cross-hairs and outside the field of view. See the next step.

3. Find the right end of the top scratch by adjusting the Y-axis of the stage, but while doing this roughly estimate how far the Y-axis must be adjusted to bring the end of the scratch into the field of view. Then rotate the culture enough to bring the end of the scratch about half-way back towards the field of view. This will bring the culture closer to the desired orientation. Adjust the stage to bring the right end of the scratch under the cross-hairs.

4. Move the stage all the way back to the left end of the scratch by adjusting only the X-axis. If the end of the scratch falls outside the field of view, repeat the above procedure using the left end of the scratch. If the left end of the scratch falls off the cross-hairs but within the field of view, rotate the dish to bring the end of the scratch half-way back to the cross-hairs, and adjust the Y-axis of the stage to bring the end of the scratch on the cross-hairs.

5. Move the X-axis of the stage all the way to the right end of the scratch. If it falls under the cross-hair, without moving the Y-axis, then the culture is properly oriented. If the cross-hairs do not fall on the scratch, repeat the process of adjustment back and forth until the left and right ends of the top scratch in the culture substratum can both be brought under the eyepiece cross-hairs by moving the X-axis of the stage without moving the Y-axis.

The next step is to establish a single reference position for the culture. By convention we use the left end of the top scratch in the culture substratum.

The stage is moved to bring the cross-hairs over this point, and the zero is set on the microscope digitizer. Then, in order to establish a reference position for each track in the left and right compartments, the position of the edge of silicone grease on each track is recorded by bringing the cross-hair over the edge of the grease and pressing the foot pedal. The grease within a single track is usually not at a perfect right angle to the measurement axis, but the resulting variation in position of the grease edge across a single track is too small to produce experimental error. As a convention the grease position farthest back towards the centre compartment is used for the track reference position to ensure that a short neurite just emerging at this point will yield a positive measurement.

Neurites are measured by bringing the cross-hairs over the farthest neurite along each track and recording this position by pressing the foot pedal. After the grease and neurite positions have been recorded, the computer calculates the neurite extension on each track by subtracting the track's grease measurement from the position of its farthest extending neurite. Then these values are used to calculate means and standard errors. The next time the same culture is measured it is oriented and the culture reference position is set in the same manner, but the track reference positions are not measured again. After neurite extension is measured, the previous track reference positions can be subtracted to give total neurite extension, or the previous neurite measurements can be subtracted to give the change in neurite extension during the intervening time.

4.3 Electrical stimulation

The silicone grease barrier through which neurites grow in compartmented cultures restricts the extracellular space, permitting extracellular electrical stimulation of the neurites. A sufficient fraction of electrical current passed between compartments crosses the neuritic membrane such that a 1 V pulse, 1 msec in duration elicits an action potential in all of the neurones assayed by intracellular recording (3). Phasic electrical stimulation at 10/sec in trains of 1 sec on and 1 sec off produces action potentials with each stimulus, but at chronic stimulation at 10/sec failures develop after a brief period, and only intermittent action potentials are eventually seen. This may reflect failure of neuritic action potentials to invade the cell body and may not indicate failures in the neurite. When cultures are subjected to stimulation in trains for several days in the incubator by means of platinum wire electrodes cemented through the dish lid, and then transferred to the intracellular recording set-up with only a brief interruption in stimulation, the neurones are observed to be firing action potentials without fail indicating that phasic stimulation over a period of days is effective. Thus, by this method the electrical activity of the neurones can be precisely controlled as they develop and grow in culture.

The technical details of electrical stimulation are relatively simple. Platinum wires are used as electrodes in each compartment and are cemented through the dish lids. Lids are reusable and are sterilized by ultraviolet light. Pulses are generated with ordinary electrophysiological stimulators, and a lab-made polarity reverser is used to automatically switch the polarity every few seconds to prevent electrode polarization. The lids can be stabilized by cementing on to them a Plexiglass bar with ends that extend beyond the dish to rest in slotted holders in a lab-made dish tray. The dish tray is wired such that small alligator clips attach to the electrodes on the dish lid and the tray connects to leads within the incubator by means of screw posts. Banana plugs should be avoided since plugging and unplugging can jostle and spill the cultures.

References

1. Campenot, R. B. (1977). *Proc. Natl Acad. Sci. USA,* **74,** 4516.
2. Schwab, M. E. and Thoenen, H. (1985). *J. Neurosci.,* **5,** 2415.
3. Campenot, R. B. (1979). In *Methods in Enzymology,* Vol. 28 (ed. W. B. Jakoby and I. H. Pastan), p. 302. Academic Press, London and New York.
4. Campenot, R. B. and Draker, D. D. (1989). *Neuron,* **3,** 733.
5. Hawrot, E. and Patterson, P. H. (1979). In *Methods in Enzymology,* Vol. 28 (ed. W. B. Jakoby and I. H. Pastan), p. 574. Academic Press, London and New York.
6. Campenot, R. B. (1982). *Devel. Biol.,* **93,** 1.
7. Campenot, R. B. (1987). *Devel. Brain Res.,* **37,** 293.

Appendix
Suppliers of specialist items

Amersham International plc, Lincoln Place, Green End, Aylesbury, Buckinghamshire HP20 2TP, UK.
Bio-Rad Laboratories, 3300 Regatta Blvd, Richmond, CA 94804, USA. (Tel. 800-227-5589)
Boehringer Mannheim GmbH, PO Box 310120, D-6800 Mannheim 31, Germany.
Calbiochem Corp., PO Box 12087, San Diego, CA 92112-4180, USA. (Tel. 800-854-3417)
Duke Scientific Corp., 1135D San Antonio Road, PO Box 50005, Palo Alto, CA 94303, USA. (Tel. 800-334-3883)
Godax Laboratories, Inc., 6 Varick Street, New York, NY 10013, USA.
Jim's Instruments, 1201 Highland Court, Iowa City, Iowa 52240, USA.
Medical Research Apparatus Corp., 1058 Cephas Road, Clearwater, FL 34625, USA.
Minnesota Datametrics, 1000 Ingerson Road, St Paul, MN 55126, USA.
Molecular Probes, PO Box 22010, Eugene, OR 97402-0414, USA. (Tel. 503-465-8300)
New England Biolabs, Inc., 32 Tozer Road, Beverly, MA 01915-5054, USA. (Tel. 800-632-5227)
New England Nuclear, 549 Albany Street, Boston, MA 02118, USA.
Pharmacia-LKB Biotechnology, 800 Centennial Avenue, PO Box 1327, Piscataway, NJ 08855-1327, USA. (Tel. 800-526-3593)
Physiologic Instruments, 7090 Miramar Road, Suite B, San Diego, CA 92121, USA.
Sigma, PO Box 14508, St Louis, MO 63178, USA. (Tel. 800-325-3010)
Soltec Co., 11684 Pendleton Street, Sun Valley, CA 91352, USA.
Stratacyte, 11099 North Torrey Pines Road, Suite 400, La Jolla, CA 92037, USA. (Tel. 800-562-8922)
Stratagene Cloning Systems, 11099 North Torrey Pines Road, La Jolla, CA 92037, USA. (Tel. 800-424-5444)
Telios Pharmaceuticals, Inc., 2909 Science Park Road, San Diego, CA 92121, USA. (Tel. 619-452-6180)
Tyler Research Instruments, 8306 Davies Road, Edmonton, AB T6E 4Y5, Canada. (Tel. 403-448-1249)
World Precision Instruments, 375 Quinnipiac Avenue, New Haven, CT 06513, USA. (Tel. 203-469-0266. Telex 493-0266 WPI UI)
Worthington Biochemical Corp., Halls Mill Road, Freehold, NJ 00728, USA. (Tel. 800-445-9603)

Index

adherens junctions (belt desmosomes) 133-9
 components 136-8
 isolation 134-5
affinity isolation of membrane proteins 41, 245, 253
aminopyridine 193
ankyrin 239, 242-3, 250
ascites fluid 39-41

biotinylating reagents, cell-surface protein labelling 234-5, 236 (*table*)

cadherins 31-53
 E- 132, 133, 227, 239
 epithelial junctional complexes maintenance 51
 N- 133
 P- 133
calcium phosphate transfection 86-8, 98-9
6-carboxyfluorescein 9
CDMA (expression vector) 57-8, 73
cell dissociation 5-6, 190
cell-cell adhesion assays 1-29
 antibody inhibition 3, 31-53
 aggregation assays 15-21
 bead (Covasphere) aggregation 18-20
 binding assays 21-9
  cell substrate adhesion assay, advantages/disadvantages 25
  cell substrate adhesion molecules (SAMs) 24-9
  single cells to cell monolayer 23
  smaller probes to cells 24
 cell aggregation 15-18
 liposome aggregation 18-20
 liposome preparation 11-12
 oligosaccharide inhibition 3
 peptide inhibition 3
 plasma membrane 18-20
 principles 2-3
 purified adhesion molcules 3, 4
 reagent preparation
  adhesion molecules coupling/labelling 11-14
  cell labelling 6-7
  coating surfaces with proteins 13-14
 reagent purification
  fibronectin receptors from placenta 9-11
  plasma membranes 7-9
 sample cell aggregation data 17 (*table*)
 smaller particles aggregation 18-20
 sorting out 21
cell-cell adhesion molecules (CAMs) 1-2, 31-53, 203, 227
 binding assays 21-4
 calcium-dependent, *see* cadherins
 calcium-independent 31
 cell membrane fractions 33-4

cDNAs coding for 75
 epithelial junctional complexes maintenance 51
 function blocking by antibody 49
 immunization procedure, forms used in 33-5
 immunoaffinity chromatography 9, 41-4
 neural cell adhesion molecule 31, 33
 purified 33
 screening procedure 36
 see also function-blocking anti-cell adhesion molecule antibodies
cell-substrate adhesion molecules (SAMs) 1, 91-109
 antibodies 27
 binding assays 24-9
 peptide sequences in active sites 27
cell-surface protein labelling with biotinylating reagents 234-5, 236 (*table*)
. Chaps 243
cingulin 125
compartmented culture analysis of nerve growth 275-98
 applications 294-8
 collagen substratum preparation 284-5
 culture construction 277-91
 culture media 291-2
 effectiveness of seal between compartments 293
 electrical stimulation 297-8
 experimental capabilities 276-7
 neurite collection 294-5
 neurite extension measurement 295-7
  alignment of microscope stage 296-7
 neuritotomy 294-5
 neurone plating 282-4
 pin rake 285-7
 silicone grease seal 293
 silicone grease syringe 289-91
 Teflon dividers 287-9
connexins 111, 112
 antibodies 159-60
 Xenopus, see under *Xenopus* oocyte
COS-7 cells 97
Covasphere
 aggregation 18-20
 proteins on 19 (*table*)
cyto-1 106-7
cytomegalovirus enhancer promoter 58
cytototactin (tenascin) 11, 28-9

deoxycholate 9, 112, 120-1
 DOC-JR fraction 123, 124
desmocalmin 132
desmocollins I/II 132
desmoglea (intracellular glue) 132
desmoglein I/II/III 132
desmoplakins 130, 205

Index

desmosomes 125–33, 203–20
  belt, see adherens junctions
  components characterization 130–2
  de novo assembly 205–20
    culture establishment 205–6
    low calcium pre-culturing 206
  fractionation (bovine tongue) 130
  isolation 126–9
desmoyokin 132
diethylaminoethyl (DEAE)-dextran transfection 57, 67
DNA
  coprecipitates, transfection of 86–8
  fragments, generation of mutations, synthesized by polymerase chain reaction 80–1
  libraries 73

epithelial cells
  electrical properties measurement 258–9
  integrity assay 51
  ion flux, see ion flux studies
  isotope flux, see isotope fluxes
  monolayer, cell–cell adhesion induction 205, 230–1
  short-circuit current technique 259–62
  transepithelial resistance measurement 261–2
expression cloning 55–73
  adaptor ligation 64–5
  antibody-coated dish preparation 69–70
  BstX1 subcloning adaptor system 63 (*fig.*)
  DEAE-dextran transfection 68–9
  panning 69–71
  poly(A)$^+$ RNA preparation from total RNA 60–1
  RNA isolation, guanidinium thiocyanate/LiCl protocol for total 59–60
  size fractionation 65
  spheroplast fusion 71–3
  SupF plasmid propagation 58–9
  vectors 57–8

Fab' fragments 14, 47–8
  ability to block function 48–51
fibronectin 9, 11, 27, 28–9, 107
  cell-spreading by 28
flow cytometry 101–3
fodrin 239, 242–3, 250
function-blocking anti-cell adhesion molecule antibodies 31–53

gap junctions 111
  biochemical analyses 111–19
  lens fibre 118–19
  myocardial 118–19
  proteins, see *Xenopus* oocyte cell–cell channel assay
  rat liver 111
    components characterization 119

distribution 111–13
functions 111–13
isolation 113, 115–19
gene
  deletion mutants at 3'-end 84
  generating mutants within 81–2, 83 (*fig.*), 93–6
gene exons removal 95
glycerol shock 98–9

halothane 188
heptanol 188
heterotypic adhesion 2
homotypic adhesion 2
hybridoma production stages 32–3
  see also monoclonal antibodies

IgG 40–1
  function blocking 48–51
  purification from serum 46–7
  subclass (mouse) 47
immunoprecipitation 38, 208–11, 245–6, 247–8
integrins 11, 91–109
  chimeric mutant 108
  genes, intron–extron structure 95
  interactions 91 (*table*)
  isolation 29
  isolation of transfectants 99–103
  live-cell staining 103–4
  mutagenesis 91–109
    applications 103–9
    assay of integrin mutants for cellular adhesion 107–8
    focal adhesion localization of mutant integrins 103–4
    localization of β_1 subunit cytoplasmic deletion mutants 104–6
    localization of integrin point mutants 106–7
  phosphorylation sites targeted 91–2
  promoters for high integrin expression level 96
  receptors 91–2
ion flux studies, epithelial 268–73
  conversion of $\mu Eq/h/cm^2$ to $\mu amp/cm^2$ 271
  non-ionic solutes, measurements for 273
  sample calculation 270
isotope fluxes, epithelial
  measurement at pre-selected voltages 272
  measurement under open-circuit conditions 272
  vibrating probe 272–3

laminin 107
liposomes 14, 15, 19 (*table*)

mannitol flux 273
Mauthner cells 112
membrane-cytoskeletal complexes analysis 239, 241, 243

Index

metabolic labelling 207–8
monoclonal antibodies 33–41
  antibody capture assays 36–7
  antigen capture assays 37–8
  antigen dose/form 33–5
  coupling to Sepharose 42–3
  functional assay 38–9
  hybridoma production stages 32–3
  inoculation routes 35–6
  limitations 38
  non-specific binding elimination 38
  screening procedure 36–9

nerve growth factor 275, 292, 293
neuronal cell culture 49–51, 275–97
non-denaturing polyacramide gel electrophoresis 246, 249–53
octanol 188, 195
oligonucleotides 93
  degenerate 96
  designing for deletions 95
  designing for point mutations 95–6
  synthetic 77

paracellular pathway, electrophysiological studies 257–73
patch-clamp analysis of gap junctional currents 167–99
  cell pairs preparation 168–70
  detection of gap junction channel currents 187–8
  direct cytoplasmic access 190–6
    attached/detached recording modes 195
    blockers 193, 195
    intracellular membranes avoidance 192
    macroscopic vs. single-channel activity 196
    optimal ionic conditions 191–2
    selectivity 193–4
    voltage dependence test 196
  double whole-cell vs. single patch approach 188–90
  dual whole-cell recording 167, 168–70
  effects of changes in junctional resistance 185–7
  equations derivation, junctional conductance measurements 171
    cellular input resistance estimation 173–5
    electrode series resistance estimation 172–3
    junctional resistance/conductance estimation 175–6
  equivalent circuit for dual whole-cell configuration 170–1
  input resistance measurements 176–7
  junctional resistance measurement 177–9
    membrane resistance effects 81
    seal resistance effects 179–81
    series resistance effects 181–3
    non-junctional resistance changes effects on R_j measurements 183–5
  sampling methods 196–9
    analytical 197–9
peptides, synthetic 33
pin rake 285–7

plakoglobin 131
plasma membranes isolation (rat liver) 113–15
plastic surfaces for adhesion experiments 13
plectin 132
polyclonal antibodies 41–8
  antigen preparation for injection 44–5
  immunizing with protein bound to nitrocellulose fragments 45
  serum treatment 46–7
polymerase chain reaction 77–84
porin OmpF 194
protein(s)
  affinity isolation of membrane proteins 41, 246, 253
  analysis of protein distributions in sucrose gradients 243
  attachment to surfaces 12–13
  complexes separated in non-denaturing polyacramide gel 250–3
    probing for membrane proteins 251
    Western (immuno-) blotting 250–1
  distribution of cellular proteins in sucrose gradients determination 244–5
  immunoprecipitation 208–11, 245–6, 247–9
  non-denaturing polyacramide gel electrophoresis 246, 249–53
  separation of protein complexes 245–53
  sucrose gradient analysis of extracted protein complexes 242–3
proteoglycan, CTB 28
proto-oncogenic tyrosine kinases 139

radioactivity counting from excised gel bands 212
radioiodination 14
radixin 138
recombinant bacteriophage antibody libraries 51–3
recombinant vectors introduction into mammalian cells 84–9
  DNA transfection by electroporation 88
  lipofection of lipoplyamine-coated plasmids 88–9
  transfection of coprecipitates of calcium phosphate/DNA 86–8

Sarkosyl (*N*-layroyl sarcosinate) 112
SDS-PAGE/fluorography 211–12
septate axons 190–1
silicone grease syringe 289–91
spheroblast fusion 57
sucrose gradient
  analysis of extracted protein complexes 242–3
  analysis of protein distributions 24
  centrifugation, protein standards 243 (*table*)
  distribution of cellular proteins determination 244–5
SupF gene 58

T cell binding to fibronectin-coated microtitre wells 27
Teflon dividers 287–9

Index

tenascin (cytotactin) 11, 28–9
tenuin 138
tetraethylammonium chloride 193
tight junction 111, 119–23, 203
   isolation of junction-enriched fraction (mouse liver) 121–3
Transfectam 88
transient expression assays 97
transient expression in mammalian cells, simian cell line COS, *see* expression cloning
tunicamycin 216

uvomorulin, functional analysis 75–89
   deletion mutants at 3'-end of gene 84
   generation of mutations in DNA fragments synthesized by polymerase chain reaction 80–1
   mutations generating within gene 81–2, 83 (fig.)
   mutations introduction by insertion of synthetic oligonucleotides 78–9
   single point mutation by polymerase chain reaction 79–80
   site-directed mutagenesis by complementary oligonucleotides 77–9
Ussing chamber 258–9, 263–6, 272

vibrating probe 272–3
vitronectin 11

Xenopus oocyte cell–cell channel assay 143–64
   connexins 158–64
      endogenous/exogenous 161–2
      eliminating exogenous 162–3
      selective inhibition of endogenous coupling 163–4
      time-pairing 162
   doping 153–5
   dual voltage-clamp 155–7
   human chorionic gonadoctropin (HCG) handling/injection 144–5
Xenopus oocyte cell–cell channel assay (*contd.*)
   independent verification of expression 158–61
      electron microscopy 160–1
      immunohistochemistry 158–60
      protein chemistry 158
   oocyte preparation 145–6
   pairing 153–4
   RNA purification 149
   mRNA microinjection 150–1
      calibration of injection volume 51
   mRNA synthesis 147–9
   in vitro transcription 147–8
   tracer flux 158
   vitelline envelope removal 151–3

ZO-1 125
ZO-2 125